SUSTAINABLE DEVELOPMENT AND MANAGEMENT OF GROUNDWATER RESOURCES IN SEMI-ARID REGION WITH SPECIAL REFERENCE TO HARD ROCKS

Proceedings of the International Groundwater Conference on

Sustainable Development and Management of Groundwater Resources in Semi-Arid Region with Special Reference to Hard Rocks

Editors

M. Thangarajan
S.N. Rai
V.S. Singh

IGC-2002
Dindigul, Tamil Nadu, India
(February 20-22, 2002)

A.A. BALKEMA PUBLISHERS   LISSE / ABINGDON / EXTON (PA) / TOKYO

Published by: A.A. Balkema, a member of Swets and Zeitlinger Publishers
www.balkema.nl and www.zsp.swets.nl

ISBN 90 5809 263 1

Printed in India

Foreword

Groundwater plays an important role in meeting the ever-increasing demands of water supply for domestic, agriculture and industrial uses. However, spurt in the industrial activities mostly during past three decades, has led to the groundwater pollution due to disposal of untreated industrial waste. Another major source of groundwater pollution are fertilizers and pesticides extensively used for agriculture. In addition to these quality related problems, indiscriminate exploitation of aquifer in many parts of the world, specially in arid and semi-arid region like India, has led to progressively declining of groundwater levels. As a result of it the farmers are forced to either deepening their wells or going for additional drilling of deeper bore wells, some times without success. All these problems point towards the need for sustainable development and management of groundwater resources.

I am indeed very happy to note that an International Groundwater Conference (IGC-2002) is being organized in a rural town, Dindigul, to share the experience and expertise of researchers and managers with the selected farmers who are really facing the water problem. I sincerely hope that the fruits of deliberations emerging out of this Conference will reach the rural folk. The pre-conference proceedings will provide information about essential data and new emerging techniques to assess the potential of groundwater resources in hard rock regions, augmentation of groundwater resources through artificial recharge, effective remedial measures to contain the migration of pollutants and the community based groundwater resources management.

I congratulate Drs. M. Thangarajan, S.N. Rai and V.S. Singh, senior scientists of the NGRI for bringing out the pre-conference proceedings in time.

V.P. Dimri,
Director, NGRI

Hyderabad-7
January 25, 2002

v

Preface

Worldwide rapid growth of population, industries and agricultural activities has brought about a steep increase in water demands which have to be met from available surface and groundwater resources. Many parts of the world have diverse hydrogeological and hydrometeorological conditions and so is the diversification in its hydrological problems which are mostly governed by the regional conditions. Vagaries of monsoon and withdrawal of groundwater in excess to replenishment of aquifer system in many parts of India and elsewhere results into continuous declining of water table causing economic problems and deterioration of water quality. The problem is manifold in hard rock regions where the water table has gone below the weathered zone and it could be extracted only from deeper fractured zones. The aquifer systems in many hard rock areas have been over exploited. At many places many farmers have drilled bore well without knowing the potential of aquifer system and thereby end up with debt trap due to failure of the bore wells. Therefore, for the proper assessment, development and management of groundwater resources to over come, or at least minimize such problems, a necessity was felt for suitable interaction among researchers, practicing hydrologists, planners and water users. For this purpose it was decided to organize an International Groundwater Conference (IGC-2002) on "Sustainable Development and Management of Groundwater Resources in Semi-arid Region with Special Reference to Hard Rocks" at Dindigul. This town was chosen as the venue of the Conference because it is suffering from scarcity of water supply due to declining of water table and groundwater pollution due to disposal of untreated effluent from farmers and other industries.

The editorial committee has received more than 180 abstracts from India and abroad which include U.K., France, Australia, Sri Lanka and Botswana, etc. and more than 120 full papers on different hydrological topics. These papers were reviewed by experts. The present volume is the collection of 70 research papers selected on the basis of the recommendation of reviewers. These papers have been grouped into following six sections:

1. Groundwater potential assessment through application of GIS, geophysical methods and remote sensing techniques

2. Groundwater recharge through natural and artificial process

3. Groundwater pollution and its remediation measures

4. Aquifer characterization of continuum and fracture media

5. Modeling of groundwater flow and mass transport

6. Community based groundwater resources management

The remaining papers are being reviewed and will be included in the second volume of the proceeding, if funds will be available. The objective of the last section is to emphasise the importance of community participation in the development and management of groundwater resources. This is an essential because they are the ultimate beneficiaries of any scheme of development and management of groundwater resources. Their participation will ensure the sustainability of any such scheme.

We hope that the present volume will cater the needs of the planners as well as users to a large extent in India and elsewhere in their effort towards better development and management of groundwater resources in a sustainable manner.

M. Thangarajan
S.N. Rai
V.S. Singh
National Geophysical Research Institute
Hyderabad

Contents

MODELLING

GROUNDWATER POLLUTION AND ITS REMEDIAL MEASURES

COMMUNITY MANAGEMENT

GROUNDWATER POTENTIAL ASSESSMENT

Intl. Conf. on Sustainable Development and Management of Groundwater Resources
in Semi-Arid Region with Special Reference to Hard Rock, (IGC 2002),
M. Thangarajan, S.N. Rai & V.S. Singh (Eds.)

Delineation of new structural controls using aeromagnetics and their relation with the occurrence of ground water in an exposed basement complex in a part of South Indian shield

Ch. Rama Rao, M.P. Lakshmi and H. V. Ram Babu
National Geophysical Research Institute, Hyderabad-500 007

Abstract

Aeromagnetic data collected over an intensely disturbed basement complex situated at the junction of two river basins viz., Marvanka basin and Chitravati basin, has been analyzed to study the nature of the relation between the drainage pattern and distribution of the aeromagnetic lineaments in an area of approximately 400sq. km in the exposed basement west of Proterozoic Cuddapah basin in South Indian shield. Interpretation of the aeromagnetic data has yielded four sets of aeromagnetic lineaments three of which have a direct and one to one correspondence with the drainage pattern and the direction of ground water movement. However, a correlative study of the magnetic lineaments in conjunction with the stream network along with the distribution pattern of 730 wells has yielded a new set of unexposed/concealed lineaments in NE-SW direction in a zone of water scarcity circumscribed by a sygmoidal shear zone. This study has indicated that the drainage pattern and the movement of ground water in this area is structurally controlled. A suggestion is made to explore for ground water resources in the NE-SW lineaments inferred from aeromagnetics rather than in the old NW-SE lineaments in this area.

Key words: Aeromagnetics, lineaments, fractures.

Introduction

Interpretation of aeromagnetic data in conjunction with hydrogeological information of intensely disturbed metamorphic basement complexes helps in delineating new structural elements which may possibly hold ground water resources. Ground water occurs in decomposed rock matrix known as the weathered zone, which ranges from 0.5m to more than 50m in thickness. Formation of a Weathered zone and its capacity to hold sufficient quantities of ground water depends on a number of geological, hydrological and lithological conditions. Ground water also occurs in the fractured weathered rock beneath the productive saprolitic layer (Jones, 1985). Arial extent of the fractures, their orientation and distribution largely depend upon the tectonic history of the area. Ground water occurrences are also related to fault and shear zone development and with dyke emplacements.

Aeromagnetic data of the exposed metamorphic basement provinces is generally characterized by high frequency magnetic anomalies with varied amplitudes. These anomalies, depending upon the tectonic history, align themselves along fractures, fault/shear zones and dyke emplacements. Abrupt termination of magnetic highs or lows, dislocations in the anomaly trends, alignment of anomalies along a line with considerable relief on either side, drag features etc., (Gay, 1972) are some of the characteristic signatures of fault/fracture lineaments and are conspicuous in many of the aeromagnetic maps of crystalline basement rocks. Aeromagnetic data has been successfully used in delineating fault/fracture controlled aquifer systems mainly in crystalline and metamorphic terrains of western Africa and other parts of the world (Astier and Paterson 1987, Bromley et al., 1994, Paterson and Bosschart, 1987). Astier and Paterson (1987) have shown, in the crystalline basement complex of West Africa, a direct correlation between the yield of bore holes and wells and their proximity to faults and dykes determined from aeromagnetic data. The influence of faults and dykes was found to be noticeable up to a distance of at least 600 m. Their studies further indicated favorable zones for ground water at the sites of fault intersections. Based on photogeological and other data, Astier (1969) used ground magnetic follow up surveys to confirm the presence of fault lineaments before pinpointing sites for resistivity surveys.

A case study to investigate the possible occurrence of ground water in a new set of unexposed lineaments inferred from the interpretation of aeromagnetic data is presented in this paper. The area of investigation forms a part of the junction of two river basins (Fig.1b) and is tectonically disturbed and lies in a zone of sigmoidal shear in the basement complex west of Cuddapah basin in the south Indian shield (Fig. 1a).

Hydrogeological setting

The area of investigation spreads over two river basins viz., Chithravati River Basin and Marvanka Basin and lies in the middle part of the junction of the two basins. The western part of the study area is occupied by granites, schists and gneissic granites where as the eastern part is mostly covered by granite gneissis and granites. Dolerite dykes traverse in ENE-WSW, NW-SE and NE-SW directions. Quartz reefs are emplaced in both the Hampapuram and Kandukuru fault/shears (Fig. 1c) whereas kimberlite intrusions are present in the central part of the area at Chigicherla. The area is intensely disturbed with repeated tectonic events. The NW-SE trending fault lineaments are older than ENE-WSW and NE-SW lineaments (Murthy et al., 1987).

Detailed geological and hydro geological investigations have been carried out in the Marvanka basin by the geological survey of India (GSI) (

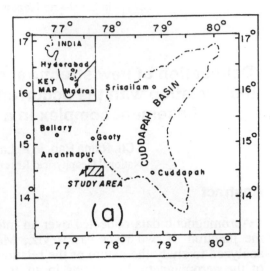

Figure 1(a): Location of study area

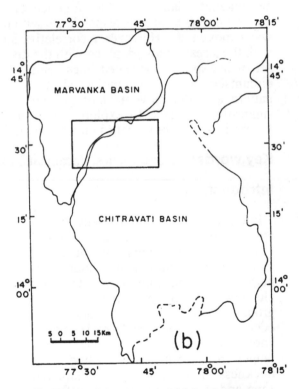

Figure 1(b): The study area and the river basins

Rama Rao 1964, 1965, 1966; Sastry 1966, Radha Krishna 1984). Raju et al., (1979) compiled detailed natural resource maps through photogeological studies and ground checks in the District of Anantpur. Raju and Murthy (1987), from a study of deep exploratory bore holes, have established the existence of potential aquifers in the fractured zones bellow a

4

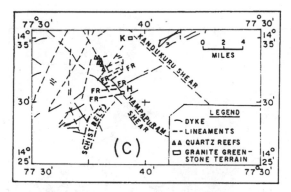

Figure 1(c): Structural elements in the study area

depth of 100m in the peninsular gneissic region of this area.

The stream network in the study area is dendritic (Fig. 2) and major lines of drainage align themselves along the directions of prominent structural lineaments. Occurrence and movement of ground water is controlled by secondary porosity caused by weathering and the existence of fractures, fault and joints. In the pediplain area, the average thickness of the weathered zone is 20m. and the average rainfall is about 560mm. The regional direction of ground water movement in western part of the area is NNE to NE and coincides with the surface water flow direction. However, in the eastern and southern parts of the area, it is ENE-WSW (Fig.3) (Chitravati Basin part) and is again controlled mainly by the geological structures. The surface water flow direction in the middle part of the area is in the NW-SE direction following the structural lineaments. However, in this case, the ground water movement seems to have been in the NE-SW direction in the zones A and B, even though the surface water direction is in the NW-SE direction.This non-preferential move-ment of ground water along the structural lineaments in this part of the area might have been due to (I) the filling up of the fault linears with highly impervious dyke intrusions and quartz reefs and/or (ii) the development of a new set of NE-SW trending lineaments as a response to the shearing and subsequent tectonic disturbance. This sort of non-

preferential movement of underground water flows, has been reported from basement regions all over the world (Proceedings of the IAH congress on Hydrogeology oh hard rocks, Oslo, 1993).

Geophysical surveys

Resistivity surveys have been extensively carried out by many governmental and non-governmental agencies in the study area for locating ground water resources to mitigate the hardship faced by the rural population in this drought-affected area. Exploration geoscientists generally follow the lineaments controlling the drainage patterns for locating ground water under the presumption that the drainage pattern is structurally controlled. Photgeological and Landsat image studies have also been undertaken to find new structural lineaments in this area (Raju et al., 1979). The Central Ground Water Board (CGWB) carried out deep drilling activity during 1984-87 to delineate deep-seated fracture-controlled aquifers in the peninsular gneissic terrain. This data has indicated, in many cases, three fracture horizons within the crustal column up to a depth of 200 meters.

The National Geophysical Research Institute, (NGRI), Hyderabad, carried out an airborne magnetometer survey over parts of Proterozoic Cuddapah basin and the adjoining crystalline basement complex during 1980-82 as a part of mineral exploration program in collaboration with the Geological Survey of India (GSI). Fig.4 is the aeromagnetic anomaly map of the study area, which has been re-interpreted for locating possible structural elements related to ground water occurrence.

Aeromagnetic interpretation

Structural interpretation map (Fig.5) has been compiled based on the general character of the aeromagnetic anomalies which include the amplitudes, wavelengths of the

5

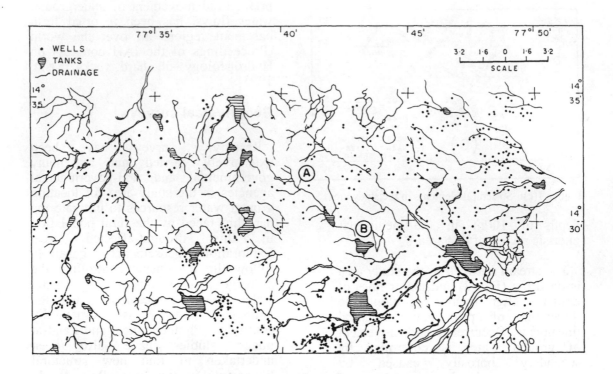

Figure 2: Map showing the locations of wells and the drainage network in the study area

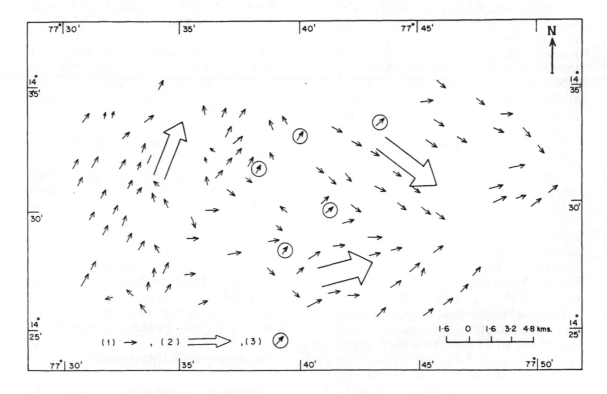

Figure 3: Water flow diagram in the study area (1) Direction of surface water flow, (2) Major structural lineaments and (3) Direction of ground water flow

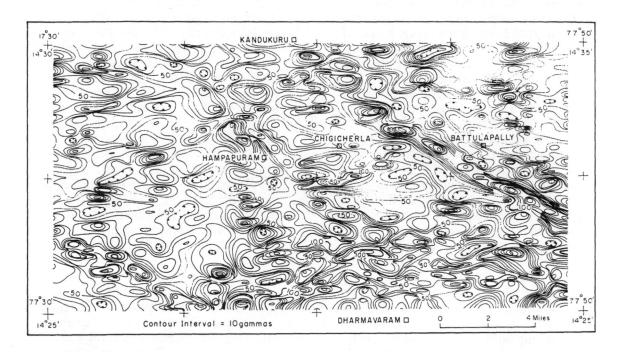

Figure 4: Aeromagnetic anomaly contour map of the study area. Contour interval 10 nT·

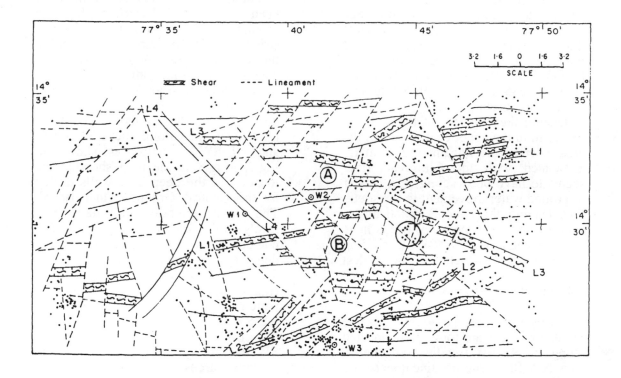

Figure 5: Map showing the locations of wells, and inferred magnetic lineaments
    L4 – Hampapuram Fault
    L3 – Kandukuru Lineament
    L2 – Chitravathi river fault/lineament

7

anomaly patterns and their orientations. This map shows four sets of inferred aeromagnetic lineaments; One set of lineaments, in the western part, traversing in NNE to NE direction, the other set of lineaments in ENE-WSW direction in the southern and eastern parts; and a third set of lineaments in NW-SE direction in the central part of the area. Comparison of these three sets of inferred magnetic lineaments with the drainage network (Fig.6) shows a one to one correspondence with each other and thus confirms that the drainage network is structurally controlled. A fourth set of aeromagnetic lineaments in NE-SW direction(Fig.5) has been inferred in the area confined by the Hamapuram-Kandakuru sygmoidal shear zone. However, there are no significant drainage streams or channels in NE-SW direction(Fig.6) coinciding with the inferred magnetic lineaments. Distribution pattern of wells and the ground water movement however indicates a NE-SW trend which may possibly show the presence of concealed fracture/fault lineaments thus coinciding with the fourth set of inferred magnetic lineaments. The ENE-WSW trending lineament L1 (Fig.5) seems to have been acting as a barrier to ground water movement, restricting the majority of the wells to the north in the upper central part of the area whereas the lineament system L2 associated with the Chitravati river seems to have been acting as a barrier to ground water movement in the south, thus leaving the zone B practically a zone of no wells. Very few wells have been located along the NW-SE trending lineament set which is the oldest lineament system in this area (Murthy et al., 1987).

However, a comparatively sizable number of wells are found to be located along the NE-SW trending lineaments. The NW-SE trending lineaments are the fractured/shear zones occupied by dolerite dykes and quartz reefs. Deep drilling data in the region of NW-SE lineaments at W1 and W2 indicated three horizons of fractures within a depth of 40m. Further drilling up to a depth of 200m did not yield either fractures or ground water. Most of the wells are confined towards the NE region of the Kandukuru lineament L3 and thus acting as a ground water barrier, leaving the areas marked as A & B as a zones of ground water scarcity. This zone is intensely fractured and sheared and tectonically disturbed with the formation of the younger NE-SW lineaments probably representing splay fault system developed due to strike slip movement along the NW-SE fault/shears and/or block rotation movement in the area (Chetty, 1995) and these may probably accommodate ground water.

Eighteen regional aeromagnetic anomalies have been interpreted using the MAGMOD software of Paterson, Grant and Waston Ltd., Canada to understand the deep structural scenario of this region. Fig.7 is the relief map of a magnetic marker horizon on which locations of 730 existing wells have been plotted. Even-though the prominent deep structural features strike in NW-SE direction following the structural grain of this part of the country, orientation pattern of the wells clearly show an ENE-WSW to NE-SW direction indicating that the main direction of the underground water movement lies along the ENE-WSW and NE-SW direction as shown in Fig.7. Fig.8 is the map showing the location of the deep bore wells drilled by the Central Ground Water Board. The data generally indicated three levels of fracture horizons. The contours presented in fig.8 indicate the depth to the deepest fracture horizon. The area marked with in the square (Fig.8) shows the general orientation pattern of the deep fracture horizon in NE-SW & ENE-WSW directions. And this corroborates the existence of NE-SW fracture system that may probably contain sources of ground water.

Conclusions

Aeromagnetic data collected over a tectonically disturbed metamorphic

8

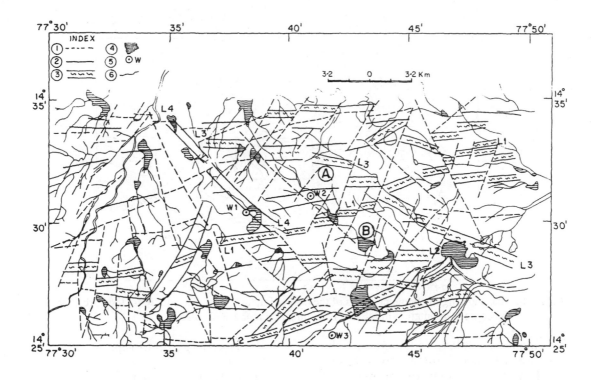

Figure 6: Map showing the correspondence between the drainage pattern and inferred magnetic lineaments. W1 & W2 are locations of deep bore wells. A & B zones of water scarcity

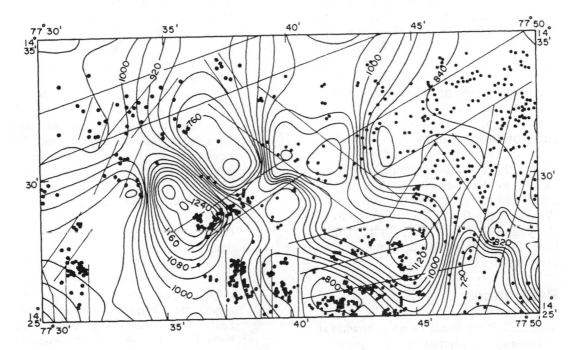

Figure 7: Map showing the depth contours of deeper magnetic horizon derived from the interpretation of regional magnetic anomalies. Contour interval 10m. Dots indicate location of wells.

9

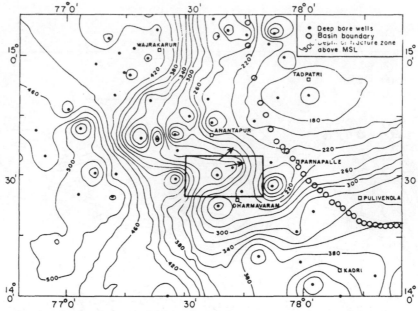

Figure 8: Map showing the depth contours of the deeper fracture zone. The square is study area. Observe the contour orientation in NE and ENE direction in the study area.

basement complex, situated at the junction of two drainage river basins has been used to identify aeromagnetic lineaments vis-à-vis their relation with the ground water occurrence. This study has brought out the fact that the inferred lineaments have one to one correspondence with the drainage network. A set of NE-SW trending lineaments has been inferred in the central part of the study area which may form a zone of possible occurrence of ground water.

Acknowledgements

We thank the Director, National Geophysical Research Institute for his kind permission to publish this work. We sincerely thank Sri. John Andrew, B,Vyagreswarudu and G.Ramachandra Rao for their neat drafting of the drawings.

References

Astier, J.L., 1969. Etude Faux souteraines au Dahoney: Interpretation des measures Geophysiques F.A.O et. Service de I Hydrauliaque.

Astier, J.L. and Paterson, N.R., 1987. Hydrogeological interest of aeromagnetic maps in crystalline and metamorphic areas.

Proceedings of Exploration' 87, pp 732-745.

Bromely, J., Mannstrom, B., Nisca, D., and Jamtild, A., 1994. Airborne Geophysics: Application to a ground water study in Bostwana. Ground Water., V.32, no.1, pp 79-90.

Chetty, T.R.K., 1995. Significance of the block rotation model in tectonics and mineralisation in Precambrian Terrains- An example from the South Indian shield. J.Geodynamics, V.20, no.3 pp 255-266.

Gay Parker, S., 1972. Fundamental Characteristics of aeromagnetic lineaments: Their geological significance and their significance to geology. Tch. Publications N0.1, American Stereo Company. Salt Lake City, Utah.

Jones, M.J., 1985. The weathered zone aquifers of the basement complex areas of Africa. Q.J. Eng.Geol. London, v.18, pp 35-46.

Murthy, Y.G.K., Babu Rao,V., Guptasama,D., Rao,J.M., Rao, M.N. and Bhattacharji,S., 1987. Tectonic petrogeological and geophgysical studies of mafic dyke swarms around the Proterozoic Cuddapah Basin, South India. Mafic Dike Swarms (eds. Halls,H.E. and Fahrig, W.F.) pp 303-316.

Paterson,N.R. and Bosschart, R.A., 1987. Airborne Geophysical Exploration for Ground Water. Ground Water, v.25, no.1, pp 41-50.

Intl. Conf. on Sustainable Development and Management of Groundwater Resources
in Semi-Arid Region with Special Reference to Hard Rock, (IGC 2002),
M. Thangarajan, S.N. Rai & V.S. Singh (Eds.)

Application of remote sensing techniques for groundwater prospects – A case study of Kaidampalli watershed, Medak district, Andhra Pradesh

S.V.B.K. Bhagavan and V. Raghu

A.P.State Remote Sensing Applications Centre (APSRAC)
DES Campus, Khairatabad, Hyderabad – 500 004, A.P.

Abstract

The present study deals with the hydrogeomorphology for ground water prospects in Kaidampalli watershed, Medak District, Andhra Pradesh. Two geological formations namely granitic gneiss and Deccan traps basalt occupy the study area. The Indian Remote Sensing satellite IRS1C/1D (PAN+LISS-III) data of rabi season is used for generation of hydrogeomorphology map on 1:25,000 scale. Well inventory data collected in the field is integrated with this thematic information and a final map is prepared. Majority of the lineaments are trending in NW-SE direction. The area is broadly divided into eight landforms of denudational origin. The ground water prospects of these units are assessed by studying various hydrogeomorphological and hydrogeological parameters.

Keywords: Watershed, Ground water, Remote sensing, Hydrogeomorphology, Rainwater harvesting

Introduction

Kaidampalli watershed is located in Alladurg Mandal of Medak District, Andhra Pradesh (Survey of India toposheet No. 56G/13 NE on 1:25,000 scale). The total geographical area of the watershed is 4034ha spread in five revenue villages namely Alladurg, Kaidampalli, Appajipalli, Reddipalli and Rampur.

The entire area is occupied by Archaean formation consisting of granites and gneisses. The Archaean rocks have suffered considerable degree of tectonic disturbances as a result of which the rocks have been metamorphosed and recrystallized. Dolerite dykes intruding the granitic rocks are common in the area. These crystalline rocks develop secondary porosity with weathering and fracturing, which enables these rocks to become water bearing and water yielding. In southwestern part of the watershed, these rocks are overlain by Basalts. Lineaments usually represent faults, fractures, shear zones, joints or an unconformity through which ground water movement takes place. These features are identified on satellite images on the basis of zonal contrast, river/stream alignment, difference in vegetation cover, knick-points in terminations of patterns etc. In these formations majority of the lineaments are along NW-SW and NE-SW directions.

The methodology adopted is remote sensing based hydrogeomorphological mapping to identify various landforms for groundwater prospects. Initially, IRS IC PAN + LISS-III merged data of the rabi season was interpreted on 1:25,000 scale to derive geomorphological information. The geological structure and tectonic phenomenon that has caused the development of secondary porosity was also identified. Information derived from remote sensing data were coupled with elevation and drainage informations obtained from the Survey of India toposheets to delineate potential zones for ground water resources development, which were further verified in the field. Field information was incorporated and final hydrogeomorphological map was prepared (Fig.1). Ground water potential zones were delineated from hydrogeomorphology, structural information and from the field well inventory data (Anon, 1999). The study area is divided into following 8 hydrogeomorphic units.

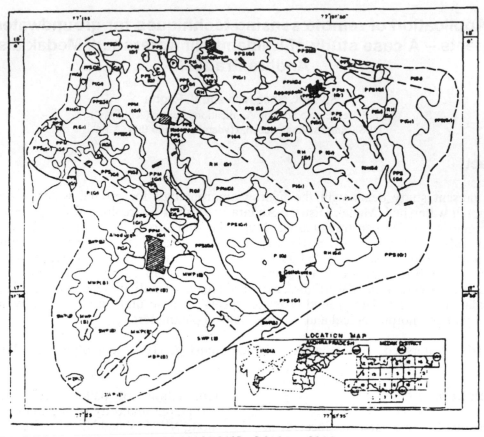

HYDROGEOMORPHOLOGICAL MAP

Figure 1: Hydrogeomorphological map of Kaidamaplli water shed, Medax district, Andhra Pradesh

- Moderately Weathered Plateau (MWP-B) - a flat and smooth surface of weathered basaltic plateau with thick black soil cover of 5 to 15 m thick weathering as observed south of Alladurg etc. In this unit, the ground water prospects are moderate to good. Good yields are expected along fracture/lineament.

- Shallow Weathered Plateau (SWP-B) - a flat and smooth surface of weathered basaltic plateau having thin black soil cover with 0 to 5 m thick weathering as observed south of Alladurg etc. The ground water prospects are poor to moderate. Moderate yields are expected along fracture/ lineaments.

- Hard Basaltic Plateau (HBP-B) - a flat, hard, and barren basaltic plateau with poor ground water prospects as observed south-west of Alladurg and south western boundary of the watershed (Fig.1).

- Moderately Weathered Pediplain (PPM-Gr) - a gently sloping, flat and smooth surface of granitic gneiss with more than 5m deep weathered material, usually, covered with red soil as seen near Appajipalli and Ramapuram. In general, the ground water prospects are moderate to good. Good yields are expected along fracture/lineament.

- Shallow weathered Pediplain (PPS-Gr) - a gently sloping, smooth, flat surface of granitic gneiss with less than 5 m deep weathered material, generally covered with red soil. This landform is observed near Appajipalli, Kaidampalli, and eastern part of watershed. In general, the ground water prospects are poor to moderate. Moderate yields are expected along fracture/lineament.

- Pediment (P-Gr) - a gently sloping, smooth surface of erosional bedrock between hill and plain of granitic gneiss with veneer of soil cover. It is observed throughout the watershed as

isolated patches. In general, the ground water prospects are poor.

- Residual Hill (RH-Gr) - a group of massive hills of granitic gneiss occupying considerably small area as observed in the east of Kaidampalli. In general, the ground water prospects are negligible.

- Dyke Ridge – a vertical to steeply dipping intrusive, often massive ridge standing above the ground level. The ground water prospects are negligible. This feature is observed in the north of Alladurg (Fig.1).

Ground Water Occurrence

The ground water occurs in the weathered and fractured rocks under water table and semi-confined to confined conditions. The area is underlain by hard rocks, which have low permeability and transmissivity values. The climate, rainfall, topography, geological setting, infiltration rate, porosity and transmissivity of the geological formations also control the occurrence of ground water. The geological and structural control play an important role in the occurrence of ground water and the yield of the wells. In this area, the ground water is generally developed by bore wells and dug-cum-bore wells.

The massive Deccan Traps or Basalts are very poor yielder of ground water. The traps with their weathering, vesicles, zeolites, cracks, fractures, intertrappean beds often form good aquifers. The wells in Deccan Traps generally have the depths ranging between 9 to 26 m. with water levels varying between 3 to 13 m. below ground level.

On the other hand crystalline formations (granites and gneisses) lack primary porosity and hence the occurrence of ground water in crystalline formations is limited to the secondary porosity developed through weathering and fracturing. Some of the dykes seem to act as barriers to the ground water movement so much that the areas of good ground

13

water potential are found towards upstream of the dykes and low potential and deep water level conditions are indicated towards its down stream of the dyke. In general, the NW-SE and E-W trending dykes act as barriers to the ground water flow and the dykes which usually follow stream courses and lineaments act as conduits for the flow of ground water.

Ground Water Recharge

In watershed planning, topmost priority must be given for providing recharge structures. Anticipated peak runoff governs the design and location of these structures. These rainwater harvesting structures include check dams, percolation tanks, farm ponds, diversion drains etc. The thematic information such as hydrogeomorphology, land use / land cover and slope are considered for selecting a suitable site for construction of a recharge structure. Keeping in view of all these data, eight check dams are recommended for recharge of ground water in the watershed area.

Conclusions

Rainfall is the only source of recharge in the dry agricultural areas. The other contributions through tanks and streams are meager in the area as majority of these water bodies are seasonal. Thus, the recharge of the ground water is less and discharge is constant. The ground water development should be made extensively in shallow and moderately weathered plateaus and pediplains of Basaltic and granitic origin. Rain water harvesting structures are to be taken up in a systematic way to augment ground water potential.

Acknowledgements

The authors express their sincere thanks to the Commissioner, Agricultural Department, Govt. of A.P. for sponsoring this project. Grateful thanks are also due to the staff members of the Agricultural Department who have helped us in completing this work.

References

Anon,1999. National watershed development project for rainfed areas (NWDPRA), Kaidampalli watershed, Medak District, Andhra Pradesh. Project Report, A.P. State Remote Sensing Applications Centre and Department of Agriculture, Government of A.P. 64p.

Bhagavan,S.V.B.K. and Raghu,V., 2001. Planning for development of rainfed areas using remote sensing and GIS techniques. In: International Conference on Remote Sensing and GIS/GPS, Hyderabad. 393-397.

WARASA,1992. National watershed development project for rainfed areas (NWDPRA), Guidelines, Department of Agriculture and Co-operation, Ministry of Agriculture, Government of India. 105p.

Intl. Conf. on Sustainable Development and Management of Groundwater Resources
in Semi-Arid Region with Special Reference to Hard Rock, (IGC 2002),
M. Thangarajan, S.N. Rai & V.S. Singh (Eds.)

Geophysical surveys for deeper aquifers in Sivaganga area, Tamil Nadu : A case study

Ramesh S. Acharya, R. Ananda Reddy, M. Rajani Kumar and K.P.R. Vittal Rao
Geophysics Division, Geological Survey of India, Hyderabad - 500068

Abstract

Geophysical surveys comprising deep resistivity soundings, gravity and magnetic surveys were carried out in Sivaganga area, a part of Cauvery sub-basin, Tamil Nadu. While the gravity and magnetic surveys along a profile between Paganeri and Natarajapuram delineated the basement configuration of the southeast sloping basin, the resistivity soundings traced multi-layered geoelectric depth section comprising of clayey-sand, sandy-clay, sand, sandstone, shale etc, with overlapping resistivities.

The resistivity characteristics of different layers were identified by correlating the lithology of a borewell with the interpreted resistivity sounding carried out in its vicinity. The sounding data enabled the identification of two aquifers viz, 1) Cuddalore formation characterized by a resistivity of 20 .m at a depth of 30-100m and 2) Gondwana sandstone characterized by a resistivity of 25 .m occurring at a depth of 100m and above, designated as shallow and deeper aquifers respectively. The chemical analysis of selected water samples led to an interesting inference that the seemingly low order of resistivities computed for aquifer zones are largely due to shale / clay intercalations rather than salinity.

Introduction

Subsurface aquifers are the main and frequently the only source of fresh water in areas where there are no perennial rivers. Sivaganga area in Tamil Nadu experiences severe drought and is included under the Drought Prone Area Programme (DPAP). Water shortage in Sivaganga area is related to various geological characteristics and structures, porous and permeable rocks of Gondwana and Cuddalore formations, flat topography and deep infiltration due to presence of laterite and alluvium. Geophysical surveys comprising deep resistivity soundings, gravity and magnetic profiling were conducted for locating deeper aquifers, because the shallow aquifers are unable to support the agricultural needs. Twenty-five water samples were collected near different soundings for chemical analysis.

The area covered lies between north latitudes 09° 45' and 10° 00' and east longitudes 78° 25' and 78° 45' covering parts of Survey of India Toposheet Nos. 58 K/5 and 9. The general gradient of this district is towards southeast with elevation ranging from 170m in the northwest to 20m in the southeast with an average gradient of 7.5 m/Km. The northern part of the district is drained by Manimuttar, Pamban and Sarugani rivers. The southern part is drained by Vaigai river. All these rivers are seasonal in nature. The area is studded with numerous rainfed tanks. The drainage pattern is dendritic in sedimentary area and dendritic to trellis in crystalline area.

Geology

Sivaganga area falls within the Cauvery sub-basin, trending NE-SW between Kallal in the NE and Manmadurai in the SW. It consists of various geological formations ranging in age from Archaean to Recent. The area is occupied by the hard crystalline rocks of Archaean age on its northern and northwestern parts (Fig.1). The crystallines are overlain by Sivaganga Formation of upper Gondwana comprising clayey-sand, sand, sandy-clay, sandstone, shale, grit and conglomerates, and Cuddalore Sandstone of Tertiary age followed by alluvium of Recent age. The formations are capped by thick laterite at places.

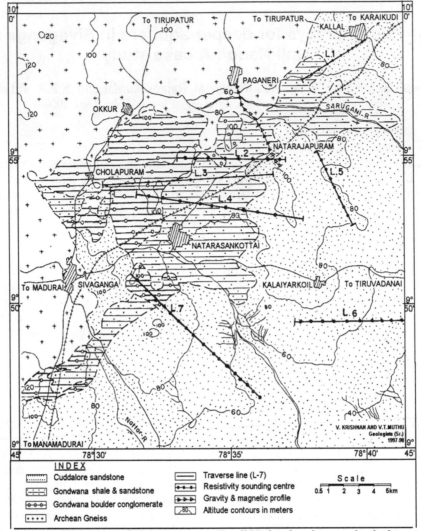

Figure 1: Geoligcal map of Sivaganga area, Tamil Nadu, showing geophysical traverses

Hydrogeology

The hydro geological condition of an area depend mainly on geology, rainfall, primary and secondary porosities of the geological units. Based on the hydrogeological setup in the Sivaganga area, the ground water conditions can be broadly described under two hydrogeological units (CGWB report, 1997).

1. Fissured hard rock aquifers belonging to Archaean age and occupying the western and north-western part of the area.

2. Porous sedimentary aquifers belonging to Mesozoic, Tertiary and Recent formations occurring in eastern and southeastern part of the district.

Hard rock terrain consists mainly of feldspathic gneisses and granitic intrusions. Ground water occurs under water table condition in general and rarely in fractured rocks at depths, at times under semi-confined conditions.

Ground water occurs at shallow depths in the Gondwana formations. The shales are impervious and constitute poor aquifers with limited ground water potential. The grits, sandstones and conglomerates are highly porous.

16

The Cuddalore formations consisting of soft grits, sandstones, clays and conglomerates constitute the potential aquifers in the area. Ground water occurs in them both under unconfined, semi confined and confined conditions.

Data Acquisition and Interpretation

Deep resistivity soundings employing Schlumberger electrode configuration with maximum half-current electrode separation of (AB/2) = 1000 - 1500m were conducted in the area employing a transmitter (TSQ-3) and a receiver unit (RDC-10). Magnetic survey was carried out employing a Fluxgate (VF) magnetometer and a Gravimeter (CG-2) was used for gravity surveys. All the instruments were manufactured by M/s Scintrex Ltd, Canada.

A total of 70 deep resistivity soundings were conducted at an interval of 500m (approximately) over seven traverse lines L-1 to L-7 (Fig.1). The details of the traverses with locations of soundings are shown in Table-1.

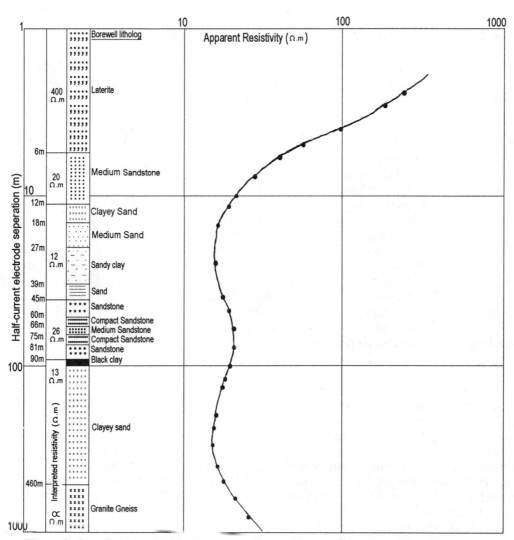

Figure 2: Contribution of Schlumberge sounding and bore well litholog, Allur, Sivaganga area

17

Table.1: Details of traverse lines with resistivity soundings-Sivaganga area.

Traverse and line Number	Direction	Length (Km)	Sounding Nos.
L-1 Kallal-Madalampatti-Muttupatti	NE-SW	5	DRS-2 to 9
L-2 Ayyampatti-Natarajapuram-Viranendalpatti	E – W	8	DRS-10 to 16
L-3 Cholapuram-Kandipatti-Singulipatti	E – W	12	DRS-17 to 18
L-4 Analmaveli-Thiruvelangudi-Ayyanarkulam	E – W	10	DRS-19 to 27
L-5 Settanendal-Tannippandal-Ottanattam	NW-SE	5	DRS-30 to 35
L-6 Paruthi-Kanmoy-Seval kanmoy-Alagapuri	E – W	7.5	DRS-37 to 46
L-7 Payyur-Allur-Oyyavandan-Pudukidi-Kallati	NW-SE	11	DRS-47 to 62

Table-2: Correlation of interpreted resistivity values with the litholog.

Bore well location & Sounding no	Observed lithology	Depth (m)	Interpreted thickness (m)	Interpreted resistivity (.m)
TWAD Board bore well, Allur village, Panangudi road, Sivaganga area, Tamil Nadu. Traverse No. L-7, Sounding No. DRS-1	Laterite	0 - 6m	6m	400 .m
	Medium Sandstone	6 - 12m	6m	20 .m
	Clayey Sand	12 - 18m		
	Medium Sand	18 - 27m		
	Sandy Clay	27 - 39m	33m	12 .m
	Sand	39 - 45m		
	Sandstone	45 - 60m		
	Compact	60 - 66m		
	Sandstone	66 - 75m	45m	26 .m
	Medium	75 - 81m		
	Sandstone	81 - 90m		
	Compact	90m continuing		13 .m
	Sandstone			
	Sandstone (Brown)			
	Black Clay			

A geoelectric model inferred from either forward or inversion technique is not unique due to the phenomenon of equivalence. Another limitation is the assumption of simplified horizontal, homogeneous and isotropic layered model. It is therefore, very essential to gather the hydro-geological information of the area and correlate it with the sounding data to achieve reliable interpretation (Dhar et al, 2000).

The Schlumberger VES curves were initially interpreted using the Master Curves published by Orellana and Mooney (1966). A standard inverse modeling program (RESIST) has been used to match the model curve with observed field data. The subsurface parameters thus obtained are correlated with the available borehole litholog.

Gravity and Magnetic (VF) observations were made at 200m intervals along Paganeri – Natarajapuram road (Fig.1), covering the Archaeans in the northwest, Gondwanas in the central portion and Cuddalores in the southeast.

Discussion of Resistivity Sounding Results

The interpretation of deep resistivity sounding (DRS-1) conducted adjacent to a bore-well drilled by TWAD Board in Allur village, on Pannangudi road, to a depth of 90m, is correlated with the lithology of the bore-well (Table-2). The sounding curve indicates a six layer setup. It is observed that laterite exhibits high resistivity (400 .m) probably due to its hardness and compact nature. Medium grained sandstone exhibit a resistivity of 20 .m. Sand, clayey-sand and sandy-clays are represented by low resistivities of the order of 12 .m. Compact and medium grained sandstone exhibit a resistivity of 26 .m which is identified as shallow aquifer. The underlying formation comprising black clay and clayey-sand are represented by resistivity of 13 .m. The resistivity sounding curve indicates bedrock at a depth of 460m (Ramesh Acharya et al., 2001).

In addition, few more lithologs of TWAD Board from different sounding locations viz., Kallal and Madalampatti over traverse L-1, Natarajapuram over L-2, Ayyanarkulam over L-4, Ottanattam over L-5 and Alagapuri over L-6 have also been correlated with the resistivity sounding curves. The comparison of sounding curves with the existing lithology of bore wells leads to the inference that, fine to medium grained sandstones exhibit the resistivities of the order of 10-20 .m and coarse grained sandstones exhibit the resistivities of the order of 20-40 .m. Coarse grained sandstone saturated with good quality of water show higher electrical resistivity as well as higher hydraulic conductivity (Choudhary et al., 1998 and 2000). The horizon with higher resistivity and thickness is likely to have more coarse grained material and thus suitable for groundwater exploitation. Based on their resistivities and possible presence of finer clastics in the former, the coarse grained sandstone is identified as

potential aquifer for ground water exploration in this area.

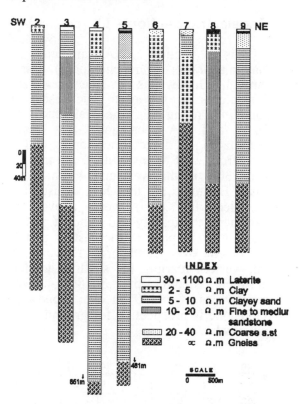

Figure 3: Interpretated vertical distribution of resistivity along with inferred geology over Traverse L-1, Sivaganga area

Geoelectric Section - L1

This section (Kallal-Madalampatti-Muttupatti) trending NE-SW direction, falls within the sedimentaries close to the contact of Archaeans in the west. The resistivity soundings along this line are multi-layered, under the category of QQH, KQKH and AAKA. The estimated depth to the bedrock varies from 150 – 650m in this section. A thick pile of fine to medium grained sandstone (10-20 .m) at a depth of 35m, having a thickness of 180m just above the bedrock is traced in one sounding (DRS-8). A conductive layer (clayey-sand) with resistivity of 5 - 10 .m, having a thickness of 150-600m is traced overlying the bedrock in DRS Nos.2, 3, 4, 5, 6 and 8. Hence there is no prospect of deeper aquifer in this section (Fig.3). The quality and yield of water in this section is likely to be poor even in shallow aquifers as the formations exhibit low resistivities perhaps representing clay / shale intercalations.

Geoelectric Section – L2

This section (Ayyampatti Natarajapuram - Viranendalpatti) passes through Gondwanas in the west and Cuddalores in the east. The resistivity soundings along this line fall into QH, QHA and HAA category. The depth section (Fig.4) depicts the sloping of bed rock from west to east with depth ranging from 30m in the west and 400m in the east. Because of the low resistivities recorded for the layers above the basement, the quality and yield of ground water may be poor in the eastern part of the profile over Cuddalores. However, in the western margin, the shallow aquifers between depth of 30–80m, with resistivities of 20–40 .m over Gondwana sandstone is prospective for ground water potential.

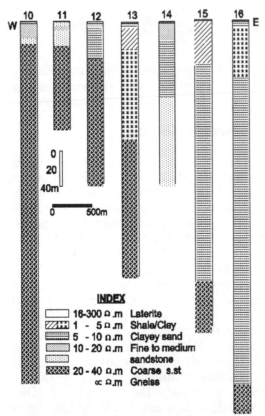

Figure 4: Interpretated vertical distribution of resistivity along with inferred geology over Traverse L-2, Sivaganga area

Geoelectric Section –L4

This section (Analmaveli - Thiruvelangud - Ayyanarkulam) cut all the formations from Gondwanas in the west and Cuddalores in the east. The resistivity soundings fall under QHA. QHAA, QHKA, KHA and AAHA category and the sounding curves exhibit the resistivities of the order of 10-20 .m over fine to medium sandstones and 20-70 .m over coarse sandstones. The depth to the bedrock varies from 380-690m in this section. A thick pile of coarse sandstone (Gondwana sandstone) at a depth of 35 to 180m with a maximum thickness of 650m was delineated in the western part of the area with possible ground water potential (Fig.5). Based on the computed resistivities and thickness for the aquifer and its inferred texture the quality and yield of ground water in the western part is likely to be good. The low order of resistivities recorded in the eastern part of the profile over Cuddalores may be due to the presence of shale / clay intercalations rather than salinity.

Interpreted resistivity distribution at 50m, 100m and 150m depth levels between traverses L-2, L-3 and L-4 (Fig.6a, b and c) indicate low order of resistivities (4–16 .m) on the eastern part of the area over Cuddalore sandstones, whereas slightly higher order of resistivities (16-60 .m) were recorded over Gondwana sandstones in the western portion of the area. It is inferred that, Cuddalore sandstones have more shale / clay intercalations than Gondwana sandstones taking into consideration their resistivity ranges. Also Gondwana sandstones are exhibiting higher resistivity variations, and are likely to be proved as a deeper potential aquifer.

Geoelectric Section – L5

Resistivity soundings over this traverse (Settanendal–Tannippandal-Ottanattam) grouped under QHKHA, QQHA, QKA, HKHA, QHA type, fall within Cuddalores. A thick pile of fine to medium sandstone with a resistivity of the order of 10-20 .m, at a depth of 40-100m with a thickness 375-525m was interpreted in

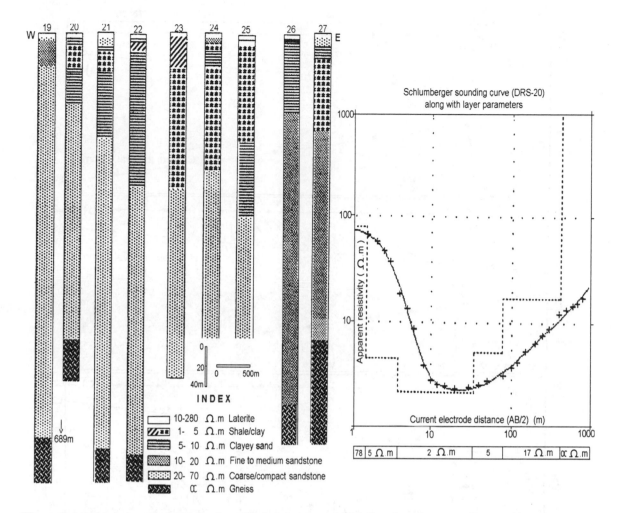

Figure 5: Interpreted vertical distribution of resistivity along with inferred geology over Tr.L-4, Sivaganga area, TamilNadu

DRS-30 to 34. Where as coarse sandstone exhibiting a resistivity of the order of 20-40 .m with a 0thickness of 350m was interpreted at 140m depth in DRS-35. The depth to bedrock varies between 400 - 600m (Fig.7). The quality and yield of groundwater in this section is predicted 'moderate' considering the ground water potential of fine to coarse grained sandstone.

Geoelectric Section – L6

This section (Paruthi-Kanmoy-Seval kanmoy-Alagapuri) also falls within Cuddalore formations, and the sounding curves are categorized under QHA, QH, HA, QH and HKA type. Comparison of litholog near Alagapuri with the interpretation of the resistivity sounding curve leads to the inference that low order of resistivities (4-10 .m) were recorded over clay/sandy clay, 10-20 .m for fine to medium sandstone and 20-40 .m for coarse sandstone. From this section (Fig.8) it is noticed that a shallow basement is encountered at 30 to 130m depth in all the soundings indicating a basement undulation / ridge like feature trending NE-SW direction between Kalaiyarkoil and Tiruvadanai. There is no deeper aquifer in this section.

21

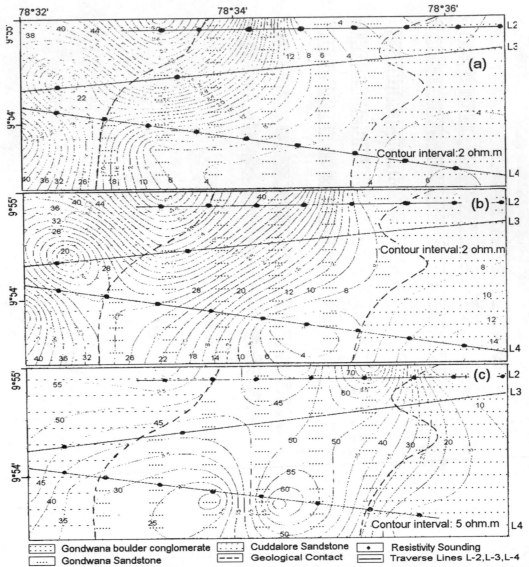

Figure 6: Interpreted resistivity distribution at a) 50m, b) 100m and c) 150 m depths between traverses L-2, L-3, L-4

Geoelectric Section –L7

Resistivity soundings in this section (Payyur–Allur-Oyyavandan Pudukidi-Kallati) grouped under HKA, AA, QHKA, HA and QKA type (Fig.9) are located over Gondwana and Cuddalore sandstones falling in NW and SE parts of the profile. Correlation of sounding curves with the existing geology from the bore well near Allur village indicate the presence of medium sand, clayey sand, sandy clay, medium/compact sandstone and black clay up to a depth of 90m. The computed depth to the deeper aquifer (Gondwana sandstone) in this section varies between

100-500m with resistivities in the range of 20-40 .m. There appears to be good potential for groundwater in the northwestern part of the profile, in view of the large thickness of aquifer zones traced in the sounding curves. Shallow aquifers (Cuddalore sandstone) with resistivity of 15-20 .m up to a depth of 100m were encountered in the southeastern part of the profile (Fig. 10). The depth to the bed rock varies from 100-850m and gradually gets shallower from NW to SE. The shallow basement encountered in the south eastern part of the traverse, was also traced in traverse L6.

22

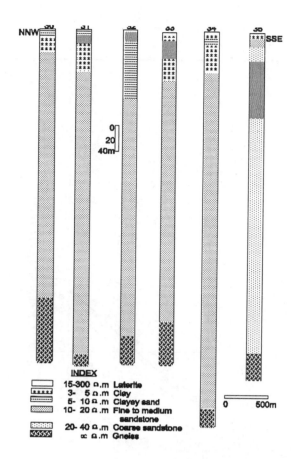

Figure 7: Interpreted vertical distribution of resistivity along with inferred geology over Tr.L-5, Sivaganga area, TamilNadu

Chemical Analysis of Water Samples

Twenty-five water samples were collected from dug / tube wells near different sounding locations for chemical analysis. The analysis data indicate that the conductivity of ground water is not uniform in the area, varying from 100 - 9000 micro mhos. The analysis of sample no.14, collected from a bore well near DRS-1, (Fig.2) over traverse L-7 gave conductivity of 388 micro mhos and total dissolved salts 213. The computed resistivity of the aquifer zone is 20 .m, corresponds to medium grained sandstone as per the litholog (Table-2) and the succeeding layer has a resistivity of 12 .m, that again corresponds to clayey-sand, medium sand, sandy clay and sand followed by layers of sandstone and clay of 26 .m and 13 .m resistivity

respectively. Hence, it is reasonable to conclude that the low resistivity (20 .m) computed for the aquifer, though of potable quality is due to clay / shale intercalations rather than salinity.

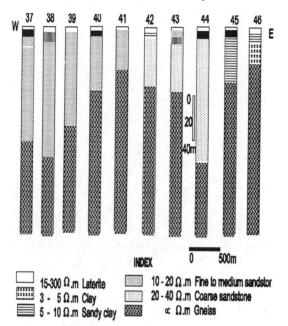

Figure 8: Interpreted vertical distribution of resistivity along with inferred geology over Tr.L-6, Sivaganga area, TamilNadu

The analysis of sample no.3 collected from an open well, over traverse L-4 gave conductivity of 409 micro mhos and total dissolved salts of 225. The resistivity of the aquifer computed from sounding data (Fig.5) is 17 .m (DRS-20) which is rather low for a water sample of potable quality. It is to be concluded that the low order of resistivities exhibited by deeper aquifers in some cases are possibly due to shale / clay intercalations rather than salinity. Hence, they could form prospective aquifer zones.

Referring to the drinking water standards (IS 10500: 1991) and the results of chemical analysis of water samples collected from different sounding locations, it is found that except the samples collected over Traverse line L-1 , which is brackish, all the remaining samples indicate that the quality of ground water is potable.

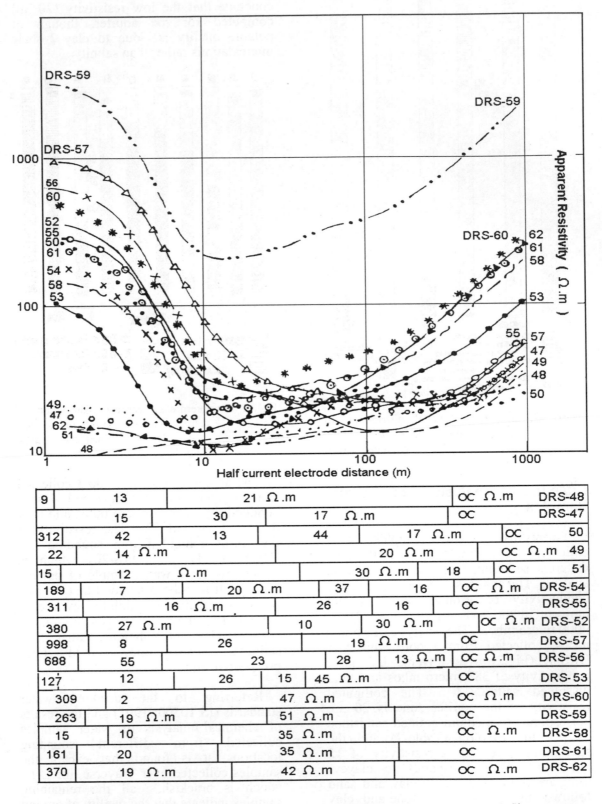

9	13	21 Ω.m				∞ Ω.m	DRS-48
	15	30	17 Ω.m			∞	DRS-47
312	42	13	44	17 Ω.m		∞	50
22	14 Ω.m		20 Ω.m			∞ Ω.m	49
15	12	Ω.m		30 Ω.m	18	∞	51
189	7	20 Ω.m	37	16		∞ Ω.m	DRS-54
311	16 Ω.m		26	16		∞	DRS-55
380	27 Ω.m		10	30 Ω.m		∞ Ω.m	DRS-52
998	8	26		19 Ω.m		∞	DRS-57
688	55	23		28	13 Ω.m	∞ Ω.m	DRS-56
127	12	26	15	45 Ω.m		∞	DRS-53
309	2	47 Ω.m				∞ Ω.m	DRS-60
263	19 Ω.m	51 Ω.m				∞	DRS-59
15	10	35 Ω.m				∞ Ω.m	DRS-58
161	20	35 Ω.m				∞	DRS-61
370	19 Ω.m	42 Ω.m				∞ Ω.m	DRS-62

Figure 9: Schlumberger sounding curves (DRS 47-62) along layer paramers over traverse L-7, Sivaganga area, Tamil Nadu

24

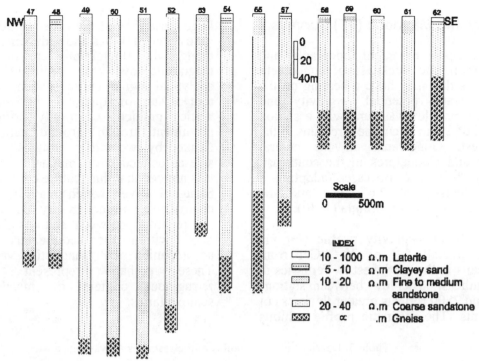

Figure 10: Interpreted vertical distribution of resistivity along with inferred geology over Tr.L-7, Sivaganga area, TamilNadu

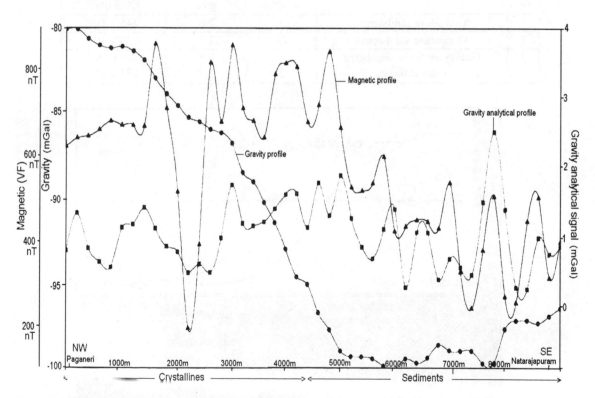

Figure 11: Gravity, Magnetic (VF) & Gravity analytical signal profiles, Paganeri-Natarajapuram road Sivaganga area, TamilNadu

25

Gravity and Magnetic (VF) Surveys

To decipher basement topography and variation in the thickness of constituent sediments that may have a bearing on the ground water potential, gravity and magnetic profile was run with a station interval of 200m to cut the Archeans in the northwest, Gondwanas in the central portion and Cuddalores in the southeast. Gravity data was processed adopting a crustal density of 2.67 g/cc and the International Gravity formula (1930).

The Bouguer gravity profile (Fig.11) records overall fall of 20 mGal from Archaeans in the northwest to Tertiaries in the southeast. The low between stations 5000 and 7800 reflects greater thickness of sediments. The magnetic profile exhibits large variations of the order of 200 – 850 nT over crystallines and sediments as well, though the fluctuations are of smaller magnitude over the sediments. The analytical signal curve (which is square root of sum of squares of horizontal and vertical gradients of gravity) brings out a prominent 'trough' around station 2500 flanked by two minor 'troughs' on either side associated with magnetic variations. It is noticed that this feature coincides with Saruguni River which may be fracture controlled. Similar gravity perturbations are recorded between stations 5000 and 8200 within the sedimentary section accompanied by magnetic variations. These perhaps represent variable ferruginous content of members of sedimentaries.

Table-3: Density and susceptibility of rock samples - Sivaganga area.

S. No	Rock sample	No.of samples	Density g/cc	Susceptibility CGS units
1	Cuddalore sandstones	3	2.0	245×10^{-6}
2	Gondwana sandstones	3	2.2	150×10^{-6}
3	Granite gneiss - weathered	2	2.5	88×10^{-6}
	- unweathered	2	2.7	195×10^{-6}

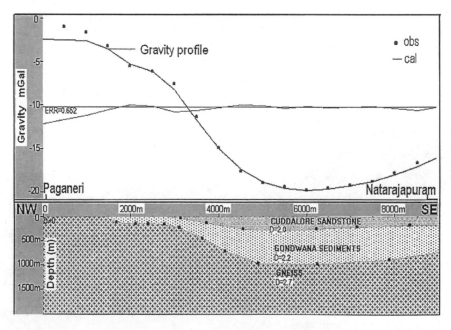

Figure 12: Gravity model along Paganeri – Natarajapuram, Sivaganga area, TamilNadu

26

The gravity profile was modeled making use of the density values of 2.0 g/cc, 2.2 g/cc and 2.7 g/cc (Table-3) for Cuddalore sandstones,Gondwana sandstones and crystallines respectively (Fig.12). The gravity low of 14 mGal between stations 4000–8000 reflects the basinal structure sloping towards southeast. Another minor gravity low, between stations 1400 and 3000 is modeled as basement undulation overlain by Gondwana sediments.

Conclusions

Resistivity sounding data revealed the existence of aquifers of large thickness in deeper sections of the Sivaganga area, forming part of Cauvery sub-basin. From the interpretation of resistivity sounding data, it is inferred that the shallow aquifers with resistivity of 20 .m occurring at a depth of 30–100m within Cuddalore sandstones and deeper aquifers with a resistivity of 20-40 .m occurring at a depth of 100m and above within Gondwana sandstones in traverses L-2, L-3, L-4, L-5 and L-7 are the prospective aquifers for potable water. It is inferred that, Cuddalore sediments appear to be clay dominated than Gondwana sandstones considering their resistivity ranges.

It is revealed that the basin is shallow in western part and gets deeper towards east. It is also revealed that a shallow basement is possible between 30-130m depth over traverse L-5 and southeastern parts of L-7, indicating a basement undulation or a ridge like feature trending NW-SE direction within the Cauvery sub-basin. From chemical analysis, it is inferred that low order of resistivities computed for aquifer zones are largely due to shale / clay intercalations rather than salinity, and the quality of ground water is potable, except over traverse L-1.

Acknowledgements

The authors are grateful to the Director General, Geological Survey of India for his kind permission to present the paper in 'International Groundwater Conference' to be held at Dindigal, Tamil Nadu, during February, 2002. They are thankful to Sri. V. Venkatachalam, Chemist (Sr.) for the useful discussions. They express their thanks to Dr. P.D. Venkateswarlu, Director (Co-ordination), Geophysics Division, GSI, SR, for the encouragement and logistic support.

References

C G W B – 1997, Ground water resources and development prospects in Sivaganga area, PMT district and adjoining areas, Tamil Nadu.

Choudhury, K., Saha, D.K., Rai, N. K., Naskar, D.C., and Ghatak, T. K., 1998, Geophysical surveys for hydro geological and esater Calcutta metropolis- Indian Minerals, Vol-51,pp:41-56.

Choudhury, K., Saha, D.K., and Ghosh, D.C., 2000, Urban Geophysical studies on the Ground water environment in parts of Gangetic Delta. Journal of Geol. Soc. of India, Vol - 55, pp: 257- 267.

Dhar, R.L., Krishna Murthy, N.S., and Prakash, B.A., 2000, Ground water Investigations in parts of Alwar District, Rajasthan. Jour. of Geol. Soc. of India. Vol - 56, pp: 151-160.

IS 10500 :1991, Bureau of Indian Standards – September, 1991, Indian Standards– Drinking water specifications (first revision) - 3rd reprint.

Ramesh S. Acharya, Rajani Kumar, M, and Ananda Reddy, R., 2001. Report on Geophysical surveys Ground water in Sivaganga dist, Tamil Nadu. G S I Un-published report,FS:1998–99.

Intl. Conf. on Sustainable Development and Management of Groundwater Resources
in Semi-Arid Region with Special Reference to Hard Rock, (IGC 2002),
M. Thangarajan, S.N. Rai & V.S. Singh (Eds.)

Application of geostatistics to the hydrochemical data from the hard rock aquifers in Sri Lank – A case study

H.A.H. Jayasena and P.T. Abeyrathne
Department of Geology, University of Peradeniya, Peradeniya, Sri Lanka
e-mail: chandra@geol.pdn.ac.lk; soiltech@ids.lk

Abstract

Geostatistical analysis was employed to hydrochemical data obtained from tube wells drilled in to hard rock aquifers in the Kurunegala district of Sri Lanka. The data set was developed using analytical results of water samples obtained from 78 tube wells and divided it into two major sub sets based on local climatological factors. The logarithmic transformations were employed for EC, TDS, HCO_3, Cl, SO_4, Ca, Mg, Na, K, TH, Alkalinity, and SAR except for pH and modeled the variograms for each parameter. Kriging were employed to variogram models exhibiting good fits with spherical, gaussian and exponential distributions and contour patterns were generated based on kriging estimates. These kriging distributions and contour patterns show general similarity except for minor variations and are controlled by morphology, lithology, fracture pattern, rainfall, and groundwater flow direction. In general TDS, EC, Ca, HCO_3, Cl, TH, and Alkalinity show very similar behavior in their spatial distribution from wetter to dryer regions of the study area. In general higher spatial continuity for the variables pH, EC, TDS, Ca, Mg lies in N30°E and west directions, however for the wetter regions it lies in the N45°W to west directions.

Keywords: Hydrochemistry, Hard Rocks, Geostatistics, Variograms, Kriging

Introduction

In many field problems, it is crucial to analyze space and time dependent data, however, at many instances these data are found to be not purely random (De Marsily, 1986). Regionalized variables such as hydrochemical data can be analyzed with the application of geostatistics (Matheron, 1965, 1971). Structure and stationarity of these data can provide implications on various environmental factors and however, even with very careful sampling and analytical procedures, sometime, it may not easy to find a conclusive picture.

The factors, which contribute to hydrochemical characterization, can be related to either the anthropogenic activities or to the water rock interaction processes. The identification of these factors is important in order to forecast their variations.

Geostatistical analysis has been widely used to identify the distribution of a set of hydrochemical and geological variables (Chambel and Almeda, 1993; de Marsily, 1986).

The Kurunegala district in the northwestern Sri Lanka with an area of 4776 km^2 was selected for the present study. It lies between the latitude 7^0 19' 00" and 8^0 10' 00" the longitude 79^0 55' 00" and 80° 35" 00" and sampling points were selected from the entire Kurunegala district (Fig. 1).

The data collected by the Water Resources Board under the Integrated Rural Development Program of the World Bank assisted groundwater project in the Kurunegala district was used (Singh and Jayasena, 1984). Hydrochemical data from 70 tube wells were selected for the present analysis based upon overall representation of the area. The variations between dry and wet zone, which has been arbitrarily demarcated in Sri Lanka, was employed in advance to separate the data set for further analysis. The following parameters have been selected for analysis: pH, EC, TDS,

In this paper we are focusing on how geostatistical analysis can be employed to hydrochemical data obtained from tube wells drilled in to hard rock aquifers.

A package GEOEAS, which is a collection of interactive software tools for performing two–dimensional geostatistical analysis of spatially distributed data was employed in the analysis (Englund and Sparks, 1988).

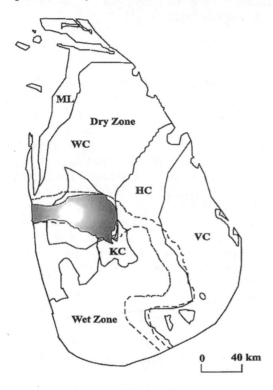

Figure 1: Map of Sri Lanka showing Kurunegala District (solid line), Major climatic zones (dashed line), Deduru Oya drainage basin (darkened) and Major geologic subdivisions of Sri Lanka (thick solid line). ML = Miocene limestone, HC = Highland Complex, KC = Kadugannawa Complex, WC = Wanni Complex, VC = Vijayan Complex.

Physiography, Geology and Hydrology

The study area is located in the northwestern province of Sri Lanka (Fig. 1), which consists mainly of Precambrian metamorphic rocks such as biotite gneiss, granite and granitic gneiss, migmatites, with minor rock formations such as quartzite,

calc gneiss, marble and other meta-sediments. In addition localized occurrences of some quartz veins and pegmatites were noted. These rocks form the Wanni and the Highland Complex rocks of Sri Lanka (Cooray, 1984; Cooray, 1992; Jayasena, 1996).

In general soil profiles in the region exhibit a thin mantle of soil over weathered rock and sound rock at the base. The thickness of the weathered superficial material in many places are restricted to less than two meters however, weathering aided by deep circulation of groundwater along structural discontinuities sometime extended this up to 15 meters (Jayasena, 1995).

Rolling topography in the northwestern lower peneplain with strike ridges and valleys on the eastern part of the study area characterize the geomorphology. The drainage network is inter-winded with the geomorphological units being more fracture controlled in the upper regions, which becomes dendretic and meandering closer to the lower western part. The study area covers several drainage basins with Deduru Oya in the middle and Maha Oya and Mi Oya flanking on either side. Scarce ground and surface water resources characterize the northern part of the area. Despite the relative scarcity, groundwater is considered as an important source of water supply for small population centers. Many private dwellers, agriculture plots, poultry and cattle farms are also using the resources.

Geostatistical analysis of data

Geoeas package is used to analyze spatially distributed hydrochemical data with parameters such as pH, EC, TDS, HCO_3, Cl, SO_4, Na, Mg, K, Ca, F/NO_3, TH, Alkalinity, SAR. At first variograms were produced (Figs. 2.1 to 2.12) and kriged to get contoured patterns (Figs. 3.1 and 3.2) in order to observe variations in the parameters. These contoured parameters were obtained based on unbiased estimation methods and hence they could be varying from those maps produced by deterministic methods (Jayasena and Dissanayake, 1995).

30

Table 1: Batch Statistics for the entire data set

P	Log	Mean	Med	S.Dev	Variance	C.Var	Skew	Kurto
pH	N	7.508	7.60	0.499	0.249	6.650	-.713	3.194
EC	Y	990.785	616.00	977.206	954931.700	98.629	2.292	8.936
TDS	Y	729.069	410.00	781.150	610195.600	107.143	2.181	7.571
HCO$_3$	Y	317.906	260.00	186.079	34625.450	58.533	.936	3.687
Cl	Y	213.164	80.00	432.827	187339.200	203.049	3.751	17.630
SO$_4$	Y	17.132	1.00	41.636	1733.581	243.028	3.400	14.385
FNO	Y	.711	0.40	1.027	1.054	144.466	2.751	10.395
Ca	Y	51.722	39.00	58.742	3450.645	113.574	3.659	19.291
Mg	Y	42.758	23.40	59.551	3546.317	139.274	3.520	17.127
Na	Y	101.682	50.00	131.986	17420.210	129.802	2.336	8.376
K	Y	8.142	3.30	14.385	206.941	176.680	3.166	12.571
TH	Y	335.796	213.00	381.273	145369.000	113.543	2.656	10.662
Alka	Y	275.284	230.00	157.242	24725.180	57.120	0.895	3.377
SAR	Y	2.369	1.60	2.247	5.049	94.829	1.753	5.702

Physical and chemical variables of the data set except temperature were initially analyzed considering the entire Kurunegala district. Subsequently the district was divided into two parts based on the climatological boundaries of Sri Lanka (Fig. 1) in order to obtain a more meaningful data distribution, and apply geostatistical analysis. The analysis consists of several steps:

a) Statistical characterization of each variable,
b) Estimation of empirical semi-variograms,
c) Fitting to the theoretical semi-variograms,
d) Direction of drift, and
e) Kriging and contouring (Interpolation of results).

Except for pH, other variables in general follow a lognormal distribution. A logarithmic transformation has thus been carried out for those variables. Statistics such as mean, median, standard deviation, variance, covariance, skewness and kurtosis for all variables were obtained (Table 1).

Variogram analysis

Variograms were generated for the data sets and kriging estimations were completed. Subsequently, new variograms were developed using semi variogram values for several directions as shown in the table 1. Given models were fitted to obtain variograms by employing different values of sill, nugget and range parameters. Every fitted model shows a nugget effect, which represents purely random noise that can be attributed to measurement errors as well as to geological variation in the examined region.

Table 2. **Directions used for variables**

Study Area	Directions	Flow Direction
Total Area	N 30° E	-
Total Area	N 45° W	-
Total Area	West	Average
Dry Zone	N 30° E	General
Dry Zone	N 45° W	-
Dry Zone	West	-
Wet Zone	N 30° E	-
Wet Zone	N 45° W	General
Wet Zone	West	

Kriging Analysis

Kriging is a weighted moving average method used to interpolate values from a sample data set on to grid of points for contouring. The kriging weights are computed from a variogram, which measures the degree of correlation among sample values in the area as a function of the distance and direction between samples (Englund and Sparks, 1988).

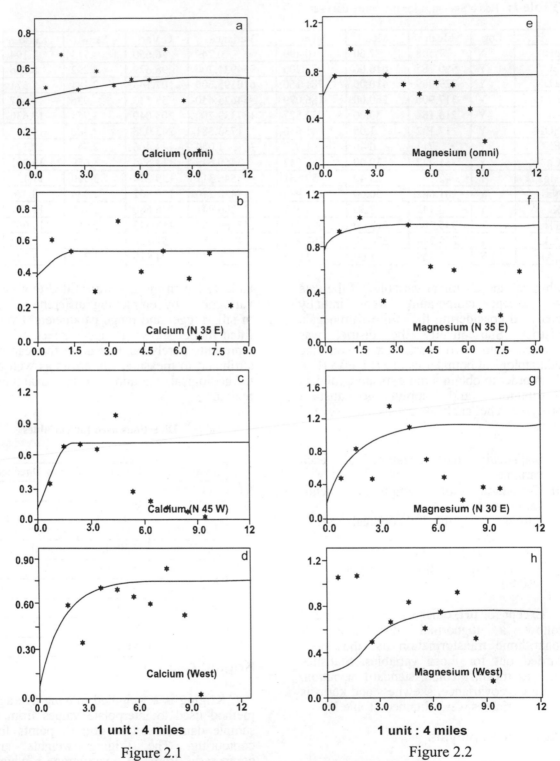

1 unit : 4 miles   **1 unit : 4 miles**

Figure 2.1   Figure 2.2

Figure 2.1: Directional semivariograms of calcium for wet zone part (r(h) on y axis, h on X axis)
Figure 2.2: Directional semivariograms of magnesium for wet zone part (r(h) on y axis, h on X axis)

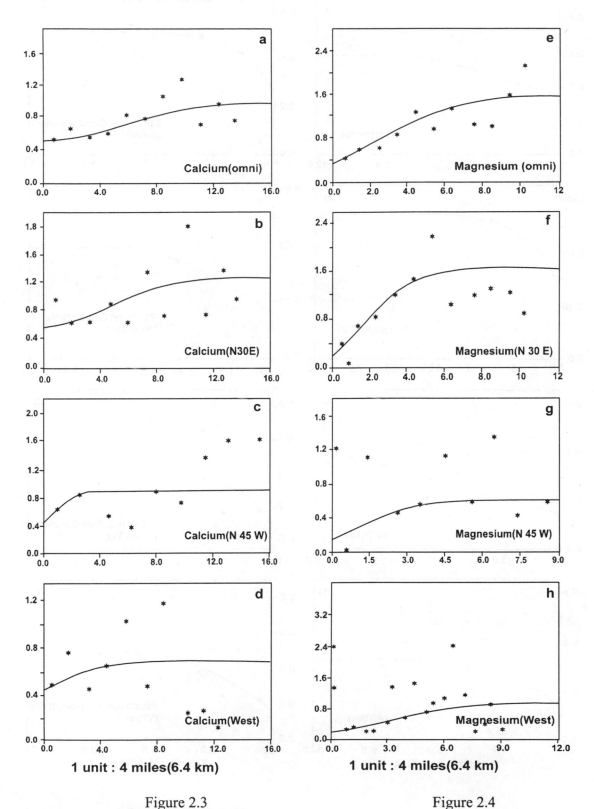

1 unit : 4 miles(6.4 km)   1 unit : 4 miles(6.4 km)

Figure 2.3   Figure 2.4

Figure 2.3: Directional semivariograms of calcium for dry zone part (ɪ(h) on y axis, h on X axis)
Figure 2.4: Directional semivariograms of magnesium for dry zone part (r(h) on y axis, h on X axis)

33

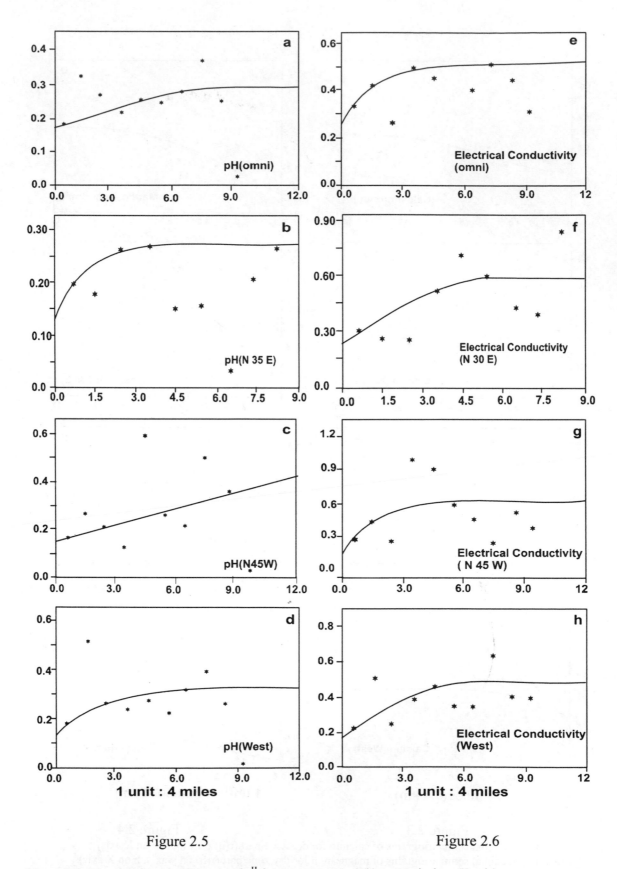

Figure 2.5

Figure 2.6

Figure 2.5: Directional semivariograms of p^H for wet zone part (r(h) on y axis, h on X axis)
Figure 2.6: Directional semivariograms of electrical conductivity for wet zone part (r(h) on y axis, h on X axis)

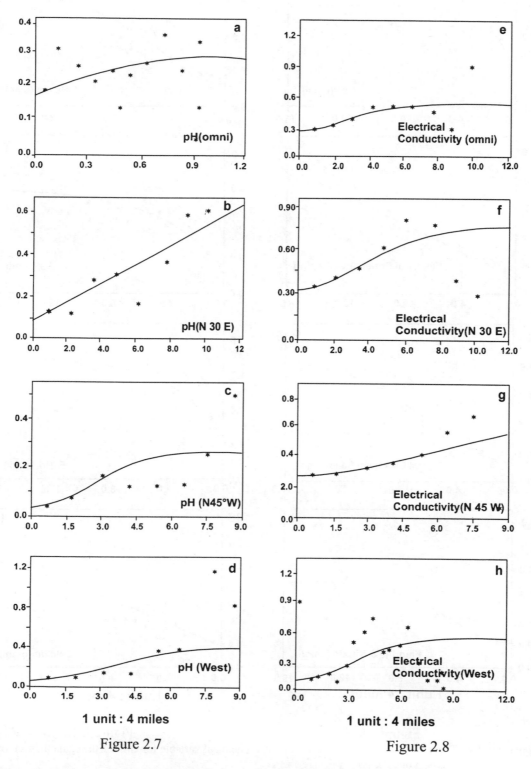

1 unit : 4 miles

Figure 2.7

1 unit : 4 miles

Figure 2.8

Figure 2.7: Directional semivariograms of p^H for dry zone part (r(h) on y axis, h on X axis)
Figure 2.8: Directional semivariograms of electrical conductivity for dry zone part (r(h) on y axis, h on X axis)

35

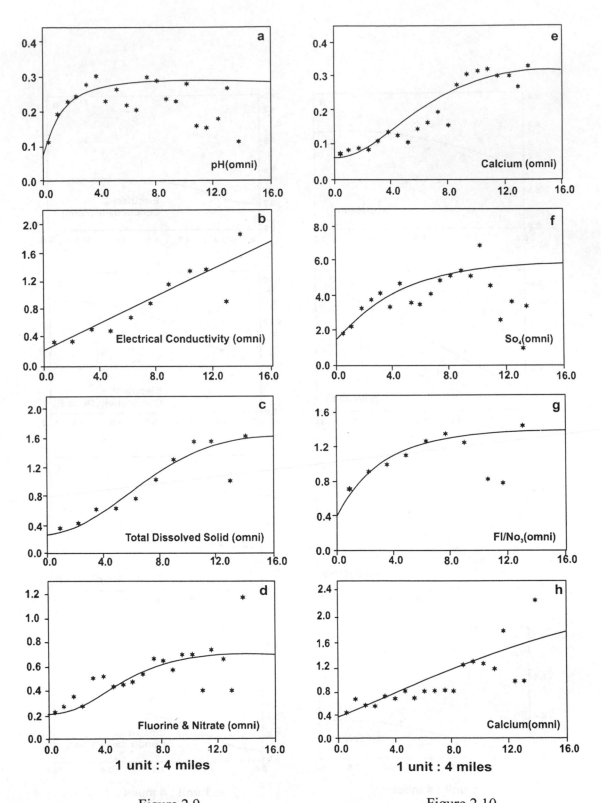

1 unit : 4 miles

1 unit : 4 miles

Figure 2.9

Figure 2.10

Figure 2.9: Non Directional semivariograms of the hydrochemical variable for entire Kurnegala district (r(h) on y axis, h on X axis)

Figure 2.10: Non Directional semivariograms of the hydrochemical variable for entire Kurnegala district (r(h) on y axis, h on X axis)

36

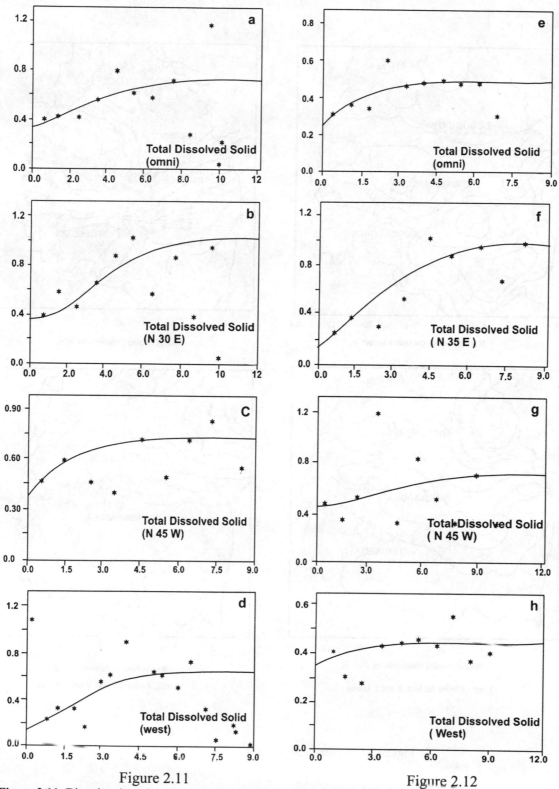

Figure 2.11

Figure 2.12

Figure 2.11: Directional semivariograms of total dissolved solid for dry zone part (r(h) on y axis, h on X axis)
Figure 2.12: Directional semivariograms of total dissolved solid for dry zone part (r(h) on y axis, h on X axis)

37

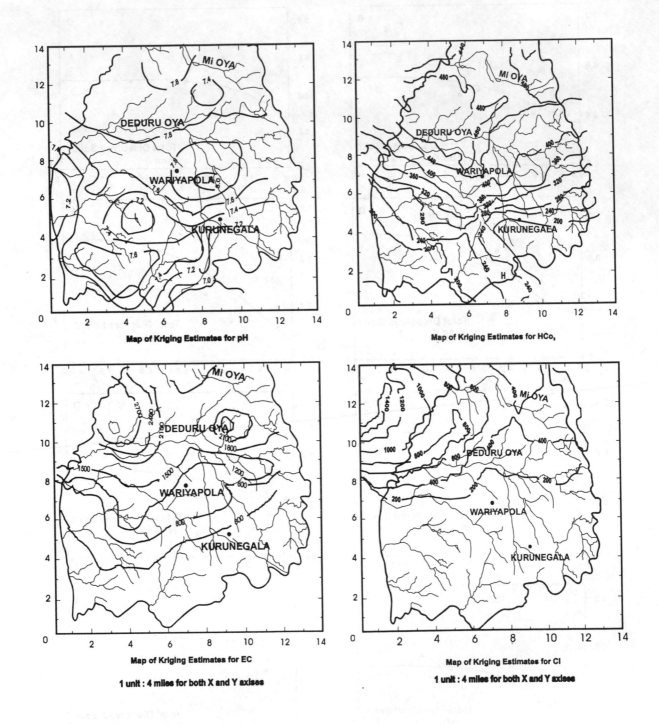

Figure 3.1: Contoured maps of Kriged estimations

38

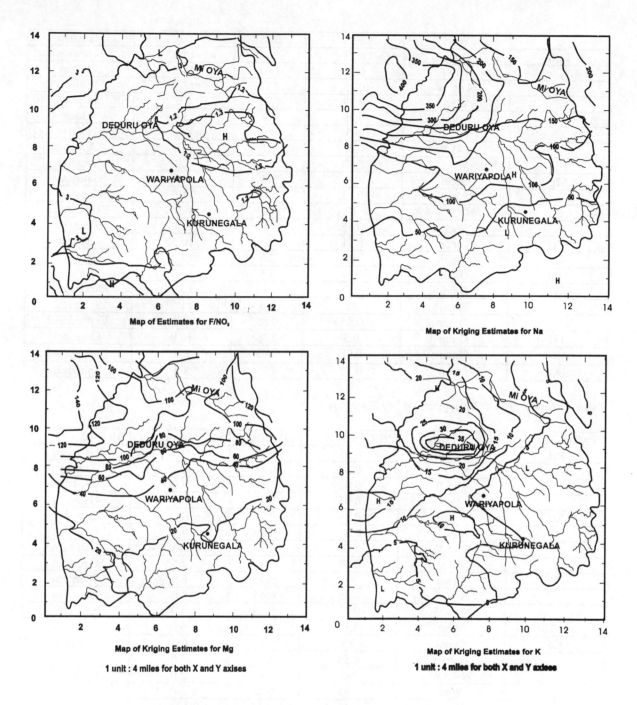

Figure 3.2: Contoured maps of Kriged estimations

Table 3: Model Data of Variograms for the Total Area

VARIABLE	DERECTION	NUGGET(c)	SILL (w)	RANGE (a)	MODEL
pH	Omni	0.07	0.22	4.5	Exponential
,,	N 30 E	0.12	0.19	4.8	Spherical
,,	N 45 W	0.1	0.175	4.9	Spherical
,,	West	0.17	0.18	4.5	Gaussian
EC	Omni	0.2	1.15	12.0	Liner
,,	N 30 E	0.27	1.1	12.0	Liner
,,	N 45 W	1.7	0.3	12.0	Liner

39

	West	0.25	0.25	6.0	Gaussian
TDS	Omni	0.28	1.4	14.0	Gaussian
,,	N 30 E	0.32	1.45	14.0	Gaussian
,,	N 45 W	0.25	2.15	12.0	Liner
,,	West	0.37	0.12	6.0	Spherical
Ca	Omni	0.38	0.9	13.0	Liner
,,	N 30 E	0.5	0.8	12.0	Gaussian
,,	N 45 W	0.45	1:1	13.0	Liner
,,	West	0.43	0.22	7.0	Gaussian
Mg	Omni	0.6	1.4	14.0	Gaussian
,,	N 30 E	0.75	1.0	11.0	Gaussian
,,	N 45 W	0	1.2	6.0	Exponential
,,	West	0.35	0.42	6.3	Spherical
HCO3	Omni	0.21	0.5	10.5	Gaussian
Cl	Omni	0.6	2.6	12.0	Gaussian
SO4	Omni	1.5	4.5	14.0	Exponential
F/NO	Omni	1.0	0.40	10.0	Exponential
Na	Omni	0.6	1.8	11.5	Spherical
K	Omni	0.45	0.86	6.0	Exponential
TH	Omni	0.5	0.75	10.0	Gaussian
Alka	Omni	0.08	0.45	8.0	Exponential
SAR	Omni	0.52	0.75	8.5	Spherical

Table 4: Model Data of Variograms of Dry Zone Part

VARIABLE	DERECTION	NUGGET(c)	SILL (w)	RANGE (a)	MODEL
PH	Omni	0.08	0.36	11.0	Gaussian
	N 30 E	0.08	0.44	10.0	Liner
	N 45 W	0.03	0.23	6.0	Gaussian
	West	0.06	0.35	8.0	Gaussian
EC	Omni	0.28	0.25	6.5	Gaussian
	N 30 E	0.31	0.46	9.0	Gaussian
	N 45 W	0.3	0.4	14.0	Gaussian
	West	0.12	0.42	7.5	Gaussian
TDS	Omni	0.0	0.95	6.0	Spherical
	N 30 E	0.35	0.65	8.0	Gaussian
	N 45 W	0.37	0.36	5.0	Exponential
	West	0.13	0.5	6.0	Spherical
Ca	Omni	0.5	0.47	10.0	Gaussian
	N 30 E	0.55	0.7	8.0	Gaussian
	N 45 W	0.45	0.45	2.0	Spherical
	West	0.45	0.23	5.0	Spherical
Mg	Omni	0.35	1.2	10.0	Spherical
	N 30 E	0.2	1.4	6.5	Spherical
	N 45 W	0.15	0.45	5.0	Spherical
	West	0.15	0.75	8.0	Gaussian

Table 5: Model Data of Variograms of Wet Zone Part

VARIABLE	DERECTION	NUGGET(c)	SILL (w)	RANGE (a)	MODEL
PH	Omni	0.16	0.12	9.0	Spherical
	N 30 E	0.13	0.145	3.4	Exponential
	N 45 W	0.151	0.2	9.0	Liner
	West	0.13	0.2	7.0	Exponential
EC	Omni	0.25	0.21	5.0	Exponential

	N 30 E	0.25	0.37	6.0	Spherical
	N 45 W	0.12	0.48	5.0	Exponential
	West	0.16	0.32	6.0	Spherical
TDS	Omni	0.25	0.25	6.0	Exponential
	N 30 E	0.15	0.8	7.0	Spherical
	N 45 W	0.45	0.20	8.0	Gaussian
	West	0.35	0.1	6.0	Exponential
Ca	Omni	0.42	0.11	8.0	Spherical
	N 30 E	0.40	0.15	1.60	Spherical
	N 45 W	0.08	0.6	2.0	Spherical
	West	0.06	0.7	4.0	Exponential
Mg	Omni	0.6	0.18	0.18	Spherical
	N 30 E	0.65	0.2	0.2	Exponential
	N 45 W	0.18	0.95	0.95	Exponential
	West	0.25	0.52	0.52	Gaussian

Table 6: Variogram model analyses for the wet zone part

Variable	Area	Direction	Anisotropy	Maximum Range	Flow Direction
pH	Dry	N 30° E	Slightly	10	General
EC	Dry	N 30° E	Moderate	9	General
TDS	Dry	N 30° E	Moderate	8	General
Ca	Dry	N 30° E	Moderate	8	General
Mg	Dry	West	Moderate	8	-

Table 7: Variogram model analyses for the wet zone part

Variable	Area	Direction	Anisotropy	Maximum Range	Flow Direction
pH	Wet	N 45° W	Moderate	9.0	General
EC	Wet	West, N 30° E	Moderate	6.0	-
TDS	Wet	West, N45°W	Moderate	8.0	General
Ca	Wet	West	Moderate	4.0	-
Mg	Wet	N 45° W	Moderate	0.95	General

Kriging yields a map of a regionalized variable, which reproduces the experimental data and provides an estimate of the variable at every other specified point. The kriged estimate is called "blue" the best linear unbiased estimate (Long and Billaux, 1987). Among all unbiased estimators, kriging estimator is the one, which produces minimum estimation variances.

The spatial pattern of the variables was obtained from point estimates in a regular grid by using kriging. We have considered a drift for the data set. The drift is based on those variables with the best fit to a trend surface and which shows significant anisotropy, i.e. logarithms of all variables except pH. For these variables, ordinary block kriging was employed. However, when ordinary kriging was employed, the sample semi-variogram estimates show biased images.

Structural Analysis

The structural analysis aims to evaluate the structure of a spatial random variable $Z(x)$ by using a probabilistic function of auto-correlation known as a semivariogram.

$$\gamma(h) = \frac{1}{N(h)}\left[Z(x_i + h) - Z(x_i)\right]^2$$

where $N(h)$ is the number of data pairs $\left[Z(x_i + h) - Z(x_i)\right]$, and h is a displacement vector. The experimental semi-variograms were also estimated in directions described in the following table (Tables 3, 4 and 5). Directions were chosen

41

considering the maximum drift and generalized groundwater flow directions of the area.

Geostatistical observations and interpretations

The tables 3, 4 and 5 show omni-directional and directional variograms with their nugget, sill and range parameters. Directional variograms were calculated only for the pH, EC, TDS, Ca and Mg.

Batch statistics were obtained using STAT-1 program of the GEOEAS. It gives basic statistics, which are useful when discussing the regional distribution of the variables. However, our main study is restricted to the geostatistical analyses based on Kriging. Block kriging of the GEOEAS package was employed for the chemical parameters; pH, EC, TDS, HCO_3, Cl, Ca, Mg, Na and TH. The resultant diagrams indicate similar behavior with small variations.

It is striking that all the parameters that were kriged show increase from wet to dry zone except HCO_3, SO_4, F/NO, Alkalinity, and SAR. This may be due to leaching of ions in the wet zone, which is much greater than that of the dry zone. The main factor that governs the leaching is precipitation. The EC, TDS, Alkalinity and Ca content increases from the wet zone to the dry zone. This can be due to low leaching of dry areas than that of wet areas.

Pockets of F and NO_3 distribution as shown in the figure (Fig. 3.2a) may represent agricultural or farming activities. Evaporation may affect to increase Cl and EC. However, EC is related with the rainfall pattern as shown in figure 3.1c where high values belong to dry segment and lower values to wet segment. Alkalinity has also been identified showing a similar trend. In general variables EC, TDS, HCO_3, Cl, Ca, Mg, Na, TH, Alkalinity behave in a similar fashion as shown in figures (Figs. 2). EC and TDS also show similar distribution pattern with a very strong correlation. The correlation coefficient of 0.976 was found for these parameters. The HCO_3 and EC have a moderate correlation of 0.566. However, EC and pH does not have very good correlation, as the correlation

coefficient is 0.101. Hence it is important to identify these relationships with the help of the sub programs Co-KRIG and XYGRAPH.

Water quality is more basic in certain areas, which may be due to anthropogenic factors such as cultivation and or structural factors such as tight fractures. When compared with TDS, both deterministic and estimated values approximately show similar behavior except for local variations. We are attempting to minimize the sample locations when estimating the kriging blocks, however, it may result in increasing the erroneous interpretations.

When considering variogram analysis of the dry zone part, it is clear that the spatial continuity of variables (pH, EC, TDS, Ca, Mg) is pronounced in the N 30° E direction. This is due to the fact that the range of the said variograms is large in the said direction for the above variables (Englund and Sparks, 1988).

The reason for this behavior may be due to the flow direction that varies from N 30° E to West. Deduru Oya, Kolamunu Oya, and Mi Oya are flowing from N30°E towards west directions, and other general drainage courses in the region are also flowing from the east to west direction. In some areas tributaries are also flowing from east and N30°E directions.

These flow directions are used in the cross validations of variogram model interpretations. Variable pH followed by TDS and Ca shows high spatial continuity among above variables, because they show higher range values. Spatial continuities of the above variables behave as the following sequence;

pH>EC >TDS, Ca, Mg.

When variogram analysis for the wet zone part is considered (Table 5), it is clear that the spatial continuity of variables pH, EC, TDS, Ca and Mg is very high in both N45°W and west directions. This may indicate that the physical observations are agreeable with the modeling results. The reason for these conditions may be assigned to groundwater flow direction, which normally varies from N45°W to west. Deduru Oya, Kolamunu Oya, and Mi Oya are flowing to N45°W to west directions, including smaller drainage causes flowing from nearly east to west direction. In some areas tributaries are flowing to west and N45°W direction. These flow directions are used in the cross validations of variogram model interpretations. The pH has high spatial continuity among the variables since it has the maximum range value. Spatial continuity of the selected variables varies according to the following sequence

pH>TDS>EC>Ca>Mg

When the total area, the dry zone part and the wet zone part are considered, nugget effect is pronounced except for TDS (omnidirection) and Mg (N45°W). The nugget effect is the result of sampling errors and short scale variable (Isaaks and Sirivastava, 1989). However all the parameters, except SO_4^{-2} and F^-/NO_3, show very low nugget value, hence the result of sampling error is minimum. The large nugget values of pH, EC and TDS indicate large short scale variability, which is commonly observed in fractured media. This short scale variability can identify with kriging estimates and also in a map of kriging estimates (de Marsily 1986). It is important to discuss directional variogram for a particular variable for selected area. For example, directional variogram of the wet zone (Figs. 2.1, 2.2, 2.5, 2.6) show following features. Calcium in the wet zone, the variogram indicates a spatial correlation (range) for distance, h, up to approximately minimum of 1.6 at N30°E and 2.0 at N45°W. Maximum values are 8 and 4 for omni-direction and west directions respectively. Tables 6 and 7 summarize maximum range observations derived from the variogram models in the dry and wet zone parts.

Anisotropy can be studied using variability of range. Several geological features can account for such an anisotropy, e.g , the lithology, the fracture pattern, and current stress field. Several methods can be employed to evaluate the anisotropy (Journal and Huijbregts, 1978; Isaaks and Sirivastava, 1989). Directional variograms were produced in order to study the anisotropy of Ca, Mg, pH, EC and TDS of the study area. By combining directional variograms with the omni directional variograms one can obtain an indication of the anisotropy. Such variograms have been developed for the two separated areas within the Kurunegala district (For the wet zone part Figs. 2.1, 2.2, 2.5, and 2.6) and (for the dry zone part Figs. 2.3, 2.4, 2.7, and 2.8)

The omni directional variogram represents a mean for all directions. Directional variograms that plot below the fitted models and does imply a longer range which may indicate directions of large spatial continuity (Englund and Sparks, 1988).

As shown in the kriged pH and TDS contour patterns (Figs 3.1a and c), groundwater flow directions in an area can be duly identified. The variations could be used even in the identification of unknown flow directions.

If the range is large for a single variable in a particular direction, it could be due to some properties of the study area. In this particular study we could identify the following properties, which may rule the differences in range.

General flow direction
Continuation of certain rock type
Relief (plane or hill)
Average maximum horizontal stresses.

pH shows minimum range value for N45°W in the wet zone and N30°E in the dry zone, which are conformable with the average flow directions as observed by previous researchers (Jayasena, 1989; Jayasena, 1993; Jayasena, 1998; Jayasena and Dissanayake, 1995).

Further detailed investigation is currently underway incorporating other parameters, which have not been discussed in the present study.

Acknowledgement

The authors are indebted to the Soil Tech Ltd. for supporting them in many ways including supply of computer facilities for the analysis. Ms. Sunethra Kanumale is mentioned for her services in drafting figures and typing the manuscript.

References

Chambel, A. and Almeida, C., 1993. Ground water hydrochemistry of the region of Portugal: Geostatistical Analysis (Eds. Sheila and David Banks), Memoirs of XXIVth IAH Congress Oslo, Norway. Section C6: 428-439.

Cooray, P.G., 1984. An introduction to the Geology of Sri Lanka. National Museum, Sri Lanka, 340pp.

Cooray, P.G., 1992, The Precambrian of Sri Lanka: A historical review, Precambrian Research, 66, 3-18.

Delhomme, J.P., 1978. Spatial variability and uncertainty in Groundwater flow parameters: A geostatistical approach. Water Resources Research 15(2): 269-280.

Englund, E. and Sparks, A., 1988. GEOEAS (Geostatistical environmental assessment software) users guide. U.S. Environmental Protection Agency, Las Vegas NV, U.S.A., 120 pp.

Isaaks, E.H. and Sirivastava, M., 1989. Applied Geostatistics. Oxford University Press New York. 28 pp.

Jayasena, H.A.H., 1989. Hydrogeology of basement complexes - A case study from the Kurunegala District of Sri Lanka. M.S. Thesis. Colorado State University, Fort Collins, CO 80525, U.S.A. 121pp.

Jayasena, H.A.H., 1993. Geological and structural significance in variation of groundwater quality in hard crystalline rocks of Sri Lanka. In Hydrogeology of Fractured rocks (Eds. Sheila and David Banks), Memoirs of XXIVth IAH Congress Oslo, Norway. Section C8: 450-471.

Jayasena, H.A.H., 1995. An analysis of fluid flow through fractured rocks. In K. Dahanayake (Editor), Handbook on Geology and Mineral Resources of Sri Lanka, Second South Asia Geological Congress Souvenir Publication, Colombo, Sri Lanka. pp 87-90.

Jayasena, H.A.H., 1996. Revision Mapping and check traverses of Wariyapola (1990), In the Geological map (1:50,000) of Wariyapola Sheet 41 (1996), Geological Survey and Mines Bureau, Colombo, Sri Lanka.

Jayasena, H.A.H., 1998. Hydrologic assessment of the Deduru Oya Basin in Sri Lanka. Multi Disciplinary International Conference on the Occasion of 50th Anniversary of Independence of Sri Lanka. University of Peradeniya, Sri Lanka. Section – G, Science and Technology 13 pp.

Jayasena, H.A.H. and Dissanayake, C.B., 1995. Analysis of hydrochemistry in the ground water flow system of a crystalline terrain. In Memoirs of XXVIth IAH Congress - Edmonton, Canada. Top in the session on Tropical Hydrochemistry. pp12

Journal, A.G. and Huijbregts, C.H., 1978. Mining Geostatistics. London. Academic Press.

Long, J.C.S. and Billaux, D.M., 1987. From field data to fracture network modeling: an example incorporating spatial structure, Water Resource .Res., 23(7), 1201-1216

Marsily, G. de, 1986. "Quantitative Hydrogeology" Academic press, 440 pp.

Matheron, G., 1965. Les variables regionalisees et leur estimation. Masson, Paris.

Matheron, G., 1971. The theory of regionalized variables and its applications. Cahiers du Center de Morphologie Mathematique , Ecole des Mines, Fontainebleau, France.

Myers, D.E. (1984), Cokriging: new developments, In "Geostatistics for natural Resources Characterization" (Verly, G., David, M., Journal, A.G., ad Marechal, A., eds), NATO-ASI, Ser. C, 122, I, 295-305. Reidel, Dordrecht, The Netherlands.

Myers, D.E. (1984), Estimation of liner combinations and cokriging, J. Int. Assoc. Math. Geol.15 (5), 633-637.

Singh, B.K. and Jayasena, H.A.H., 1984. Hydrogeology, exploratory drilling and groundwater resources potential in the Kurunegala District. Water Resources Board Report, Colombo, Sri Lanka., 293 pp.

Warrick and Myers, 1987. Kriging in the hydroscinces. Adv. Water Resour. 1:251-266.

Application of magnetic, HLEM and VES techniques for delineating fractured bedrock aquifers in unexposed areas – A few case studies from southeastern Botswana

H. Kumar[1], F. Linn[1] and I. Mannathoko[2]

1 Water Resources Consultants (Pty) Ltd., P.O.Box 40459, Gaborone, Botswana.
2 Department of Water Affairs, Pvt. Bag 0029, Gaborone, Botswana.

Abstract

Groundwater is a crucial resource in Botswana (Southern Africa) where surface water supplies are extremely limited. In much of the country, aquifers are developed in fractured sedimentary, volcanic and granitic terrains. However, delineation of fractured bedrock aquifers is often difficult due to lack of surface exposures as most of the country is covered with variably thick Kalahari Beds (mainly sand, calcrete, silcrete and clay). As a consequence, drilling success rates are poor when more basic methods like aerial photo interpretation and magnetic profiling are used alone. The recent experience in Botswana has shown that comprehensive siting programmes including ground reconnaissance, aerial photo/remote sensing data interpretation and, integrated geophysical survey greatly improve drilling success rates. Integrated geophysical survey comprised of Magnetic, Horizontal Loop Electro-Magnetic (HLEM) profiling and Vertical Electrical Soundings (VES) techniques has proven to be effective methodology in delineating fractured zones and siting boreholes in such areas. A few case studies are presented from four villages in the southeastern Botswana, underlain by different geologic units. The methodology applied has successfully delineated fracture zones associated with lithological contacts, faults/shears and kimberlite intrusions. The study indicates that magnetic and HLEM profiling in conjunction with imagery analysis and ground reconnaissance is extremely useful in precise delineation of fracture zones and siting successful boreholes.

Key words: Fractured aquifers, borehole siting, magnetic, HLEM, and VES

Introduction

One of the major challenges in developing groundwater resources in Botswana is the delineation of fractures zones due to presence of Kalahari Beds (Tertiary and Quaternary Age) covering almost 85% of county. Application of integrated magnetic and HLEM profiling with VES techniques in conjunction with aerial photographs and aeromagnetic interpretation are becoming established as an effective approach for delineating fracture zones in most of the country.

Following this approach, a comprehensive study of landsat image/aerial photographs, aeromagnetic image and existing hydrogeological information is conducted on a regional scale. It provides a base to focus on particular areas with expected structural features of groundwater interest, such as faults/ shear zones, lithological contacts and intrusions. These features are then delineated on the ground with magnetic and HLEM profiling techniques by taking measurements along several profile lines across the expected structural feature. Due to contrast in the physical properties (conductivity, magnetic susceptibility) across these features, they become apparent as anomalous zones on geophysical profile line plots. Correlation of these anomalies with existing hydrogeological information allows assessment of the degree of development of the fracture zones. Well-developed fracture zones present at promising depth are further investigated with VES technique to estimate thickness of weathered and fractured zones.

This paper illustrates four case studies from a project carried out under the auspices of the Department of Water Affairs (DWA), Botswana, to develop groundwater sources to meet future water supply of four villages (Figure 1), Dikgonnye, Gasita, Mabalane and Mogonnye, located in southeastern Botswana. The study showed that integrated magnetic and HLEM profiling with landsat image/ Arial-photographs is extremely effective in delineating vertical and sub-vertical fracture bedrock. In the cases described in this paper, it is clear that fracture zones would have been missed if only one of these techniques was applied.

Application of VES was limited in few cases because of several factors such as (1) the thick sequence of highly resistive dry sands (Kalahari Beds) hampering signal penetration, (2) insignificant resistivity contrast between thin fractures and solid rocks, and (3) failure of a homogenous stratified earth model in these conditions.

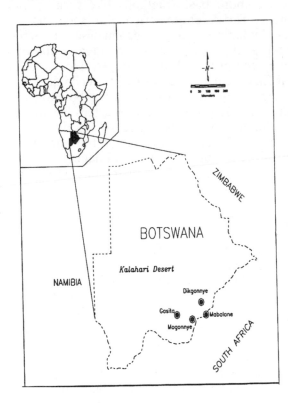

Figure 1: Location map

Geophysical Investigations

Approach

The target in all of the exploration areas described below was fractured aquifers, primarily associated with structural

features. It was considered that fracturing in bedrock is most significant in the vicinity of various structural features like faults, lithological contacts and igneous intrusions. Zones with large aerial extent can be identified from aerial photographs/ landsat imagery, interpreted aeromagnetic data and ground geophysical surveys. Due to presence of Kalahari Beds cover in much of the area, the precise location of fracture zones was difficult from aerial photographs/ landsat imagery and aeromagnetic interpretation alone. Therefore application of ground geophysical surveys was necessary with the aim of precise delineation of fracture zones, and estimating the depth and thickness of these zones.

Several structural features were identified from aerial photo/ landsat imagery lineaments, aeromagnetic data (4 km line spacing) and existing geological maps. To verify the location of these features, geophysical profile lines were setup to cross them.

Magnetic profiling and HLEM profiling in conjunction with VES techniques were considered appropriate for the study. Magnetic profiling is the most effective method in demarcating faults, lithological contacts and intrusions whereas HLEM is effective in demarcating vertical and sub-vertical fracture zones. The application of both techniques was adopted to overcome ambiguities in the interpretation of each individual method. For example fracture zones by themselves do not commonly have a clear magnetic signature but when magnetic anomaly is correlated with HLEM anomaly at same location can characterise fractures zones and associated structural features. Following the profiling, VES technique provides an estimate of thickness of weathered and fractures as well as depth to the basement.

Equipment Used

A Gemlink GSM-19GW proton precession magnetometer at 20-meter reading intervals was used for magnetic profiling in all the case studies discussed in this paper. This equipment measures Earth's total magnetic field.

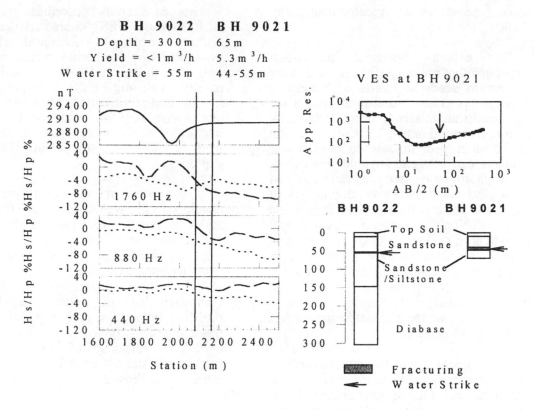

Figure 2: BH 9022/9021, Geophysis & drilling results

Following the magnetic profiling, HLEM profiling was carried out along same profile lines. An Apex Max-Min-I-10 EM unit at 20-meter reading interval was used for HLEM profiling. Two components; Inphase (IP) and Outphase (OP), of secondary fields were measured at each station using three base frequencies, selected among 7040, 3520, 1760, 880 and 440 Hz. A coil separation of 100m coil was selected in all case studies. The depth of investigation is generally considered as half of the coil spacing, however it depends on several other factors like, frequency and resistivity of different subsurface formations.

VES were conducted at selected locations identified from magnetic and HLEM profiling. A D.C. Resistivity Meter was used in Schlumberger electrode array with current electrode spacing (AB) ranging between 500 and 800m.

Magnetic and HLEM data was analysed qualitatively using Jandel Scientific software and VES data was interpreted using Resix-Plus software.

Results and Discussion

Case Studies

(1) Dikgonnye

This case study describes magnetic, HLEM and VES characteristics and drilling results in fractured sandstone associated with diabase intrusion.

This study area is underlain by sedimentary units of Waterberg (Middle Proterozoic) and Karoo (Later Paleozoic – Early Mesozoic) Supergroup and intruded by diabase (post-Waterberg – pre-Karoo). Waterberg Supergroup units are primarily sandstone, conglomerate, shales and siltstones with the overlying Karoo Supergroup characterised by feldspatic sandstone, mudstone and shales (Key, 1983). Diabase intrusions are reported to be 10 to 40m thick in the area while at some places they are greater than 200m thick.

The existing borehole information suggested that main aquifers in the area are present in sandstone units of Waterberg Supergroup. These sandstone units have enhanced transmissivity if connected with a large fractured network mainly occurring in the vicinity of diabase contacts. These contacts can be generally identified from aerial photographs, aeromagnetic data and geological maps. However, due to lack of surface exposures their precise delineation was difficult. Ground geophysical investigations comprising of magnetic and HLEM profiling across these features helped in precisely delineating fracture zones and siting boreholes. A typical magnetic and HLEM profiling response, in the area, at diabase and sandstone contact is shown in Figure 2.

Lithological contacts between diabase and Waterberg sandstone is clearly seen on Magnetic and HLEM profiling. A lower magnetic response and relatively negative IP and OP response towards the progressive side of profile line represents Waterberg sandstone whereas higher magnetic response and higher IP and OP on the other side represents presence of diabase intrusion. Two boreholes; 9021 and 9022, were subsequently drilled in the fractured zones interpreted from magnetic and HLEM anomaly. Interpreted depth of weathered and fractured zone (10-75 Ωm) from VES (at 9021) was 61m, no VES was conducted at 9022. Borehole 9021 encountered significant fracturing in sandstone between 44 to 55m, whereas 9022 encountered minor fracturing at 55m. Estimated yields of 9021 and 9022 were 5.3m^3/h and <1m^3/h respectively. The variations in yields of both boreholes are correlated with varying degree of fracturing. A less developed OP anomaly at 9022 suggests that this borehole is located at less developed fracture zone. The study suggested that all applied geophysical methods were successful in delineating fracture zones.

(2) Gasita

This case study describes magnetic, HLEM and VES characteristics and drilling results of fractured rhyolite associated with kimberlite intrusion. This study area (Carney, 1994) is underlain by mainly volcanic units comprised of rhyolite and porphyritic felsites (Archean – late Archean). Occurrence of kimberlite intrusions was also reported in the area from a previous study for diamonds exploration.

Aquifers in volcanics are well developed in weathered and fractured zones in felsites associated with several lineaments and kimberlite intrusions (pipes). These features were identified from aeromagnetic data (Prakla - Seismos, 1987). Ground geophysical surveys comprised of magnetic and HLEM profiling helped in precisely delineating these pipes. A typical magnetic, HLEM and VES response of weathered and fractured felsite associated with kimberlite intrusions is illustrated in Figure 3.

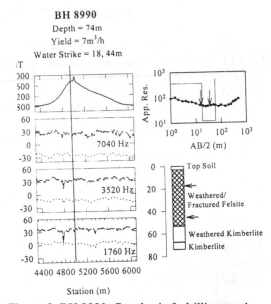

Figure 3: BH 8990, Geophysis & drilling results

The location of a kimberlite pipe was inferred at the peak of high magnetic response (Figure 3). The HLEM anomaly however was insignificant probably due to insufficient conductivity contrast between kimberlite intrusion and country rock (felsite). VES interpretation suggested weathered and fractured zones between 18 and 60m as seen from a layer of 12 Ωm resistivity. A borehole, 8990, drilled at the site penetrated weathered kimberlite at 55m depth followed by dense kimberlite at 70m. A water strike (7 m³/h) was encountered in weathered felsite at 44m depth.

Another borehole, 9046, was drilled on a similar magnetic response (Figure 4). Here, the kimberlite was not encountered. However, highly weathered and fractured felsite was encountered between 7 and 56m. The main water strike (10m³/h) was encountered at 49m depth in weathered felsite.

The study suggested that aquifers associated with kimberlite intrusions in the area have significant groundwater potential and can be identified primarily due to high magnetic response. Quantitative interpretation of individual anomaly may help in estimating geometry of these pipes.

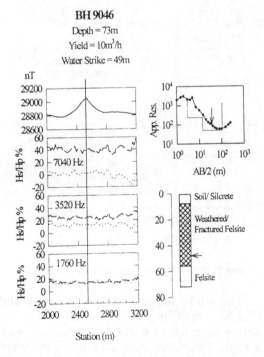

Figure 4: BH 9046, Geophysis & drilling results

(3) Mabalane

In this case study magnetic, HLEM and VES characteristics and drilling results of aquifers associated with a faulted contact

between granitiod gneiss and schists with volcanic andesitic basalt are discussed.

This study area is dominantly underlain by gneiss and granitoids (Archean Age) with some andesitic and basaltic intrusions (Archean Age) (Carney, 1994).

Previous drilling indicated that significant aquifers in the area are mainly confined in fractured gneiss and basalts associated with faults and contact zones. As such, geophysical surveys comprised of magnetic, HLEM and VES was carried out across several lineaments, faults, lithological contacts identified from aerial photographs, geological maps and aeromagnetic data. A typical magnetic, HLEM and VES response of a fault/ contact between gneiss and basalt is illustrated in Figure 5.

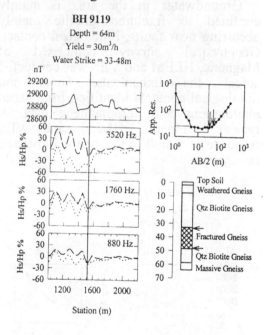

Figure 5: BH 9119, Geophysis & drilling results

The location of the fault/ contact is clearly demarcated by HLEM profiling as seen from change in regional IP and OP response. Smooth IP and OP responses towards the progressive side of profile line represents basalt. Variation in magnetic field is insignificant across the structure. VES interpretation suggested fracturing between 7 and 15m as indicated by a layer of 27 Ωm resistivity. A borehole, 9119, drilled on HLEM anomaly encountered significant fracturing in gneiss between 33 and 48m. This borehole was the highest yielding borehole in the area and test pumped for 72 hours at a discharge rate of 30 m^3/h. The study suggested that HLEM profiling was most effective in delineating contact between gneiss and basalt. Magnetic properties of both rocks do not have significant variations. Fracturing below 15m was not resolved with VES.

(4) Mogonnye

At this site magnetic, HLEM and VES characteristics and drilling results of a fault/ contact between dolerite and granite are discussed. This study area is underlain by granite (Archean Age) with dominantly intruded by felsic tuffs (late Archean Age) and dolerite sills (Key, 1993).

Groundwater in the area is mainly confined to fractured granite mainly occurring near faults, lithological contact. Geophysical survey comprised of Magnetic, HLEM and VES was conducted across the lineaments, faults and lithological contacts identified from aerial photographs and aeromagnetic data. A typical magnetic, HLEM and VES response of a fault in the area is illustrated in Figure 6.

A fault is delineated from magnetic and HLEM profiling near 1000m (Figure 6). The investigated depth from VES at this site was 14m. A borehole, 9074 (46m deep) drilled on magnetic and HLEM anomaly. Borehole penetrated weathered dolerite between 10 and 30m followed by granite. A water strike (6m^3/h) was encountered at fractured dolerite and granite contact at 29m. The borehole was subsequently tested for 72 hours with a discharge rate of 4m^3/h. An older borehole, 1277, located in this study area has one-fifth (1.2m^3/h) of the yields of 9074. Magnetic and HLEM profile response (Figure 7) suggested that borehole 1277 is located at the edge of a significant magnetic and HLEM anomaly.

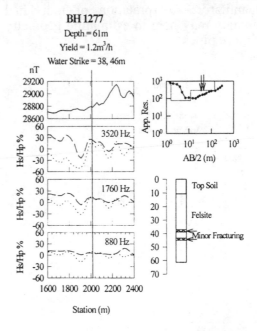

Figure 7: BH 1277, Geophysis & drilling results

The study suggested that magnetic and HLEM profiling were effective in delineating fracture zones associated with a fault. Fractured zones below 14m were not resolved with VES.

Conclusions

Groundwater exploration and development is often difficult in Botswana due to the complexity of the fractured aquifers and the ubiquitous cover of

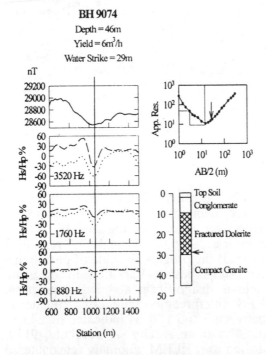

Figure 6: BH 9074, Geophysis & drilling results

Kalahari Beds, which obscure bedrock exposure and limit more traditional exploration methods such as interpretation of imagery (air photos, satellite) and field mapping of structures. In the four case studies that were presented in this paper, the additional use of integrated geophysical methods allowed increased success rates in siting of productive water supply boreholes. In the four examples cited, the geophysical methods employed were magnetic and HLEM profiling, and VES. These methods were used in conjunction with interpretation of satellite/aerial photograph imagery and analysis of existing hydrogeologic data sets. In each of the cases, representing different geological formations, subsequent drilling success rates were achieved as a result of the comprehensive siting programme employed. These improved success rates in such areas (unexposed) are considered to outweigh the additional cost and time expenditure required to undertake the integrated geophysical surveys.

References

Carney, J.N., Aldiss, D.T. and Lock, N.P., 1994, The Geology of Botswana, Bulletin 37, Geological Survey Department, Lobatse, Botswana.

Key, R.M., 1983, The Geology of the Area Around Gaborone and Lobatse, Kweneng, Kgatleng, Southern and Southeast Districts, District Memoir 5, Geological Survey Department, Lobatse, Botswana.

Prakla-Seismos AG, 1987, Airborne Magnetic Survey of Eastern Botswana, Geological Survey Department, Botswana. Interpretation Sheet 2426A and 2425B (1:125,000).

Intl. Conf. on Sustainable Development and Management of Groundwater Resources
in Semi-Arid Region with Special Reference to Hard Rock, (IGC 2002),
M. Thangarajan, S.N. Rai & V.S. Singh (Eds.)

Structurally disturbed terrain delineated by electrical resistivity method for groundwater exploration

T. Venkateswara Rao

National Geophysical Research Institute, Hyderabad-500 007 (A.P.), India
e-mail: postmast@csngri.ren.nic.in

Abstract

Groundwater movement is controlled by geological structures like faults, shear zones and lineaments, which play a vital role in tectonically disturbed areas. The Gani-Kalva-Veldurthi fault in Kurnool district (A.P.) is a result of the tectonic movements caused in the Cuddapah basin. The area comprises Gulcheru quartzites and Vempalle limestones. An integrated approach viz., geomorphologic, hydro geological and electrical resistivity investigations were successfully adopted to study and to demarcate the effect of the hydrogeological features on the groundwater regime in the study area. Detailed dipole-dipole and three-electrode array profiling and vertical electrical sounding (VES) of Schlumberger array were carried out in and around the geologically established Gani-Kalva-Veldurthi fault, near Veldurthi village, which is situated on topographically elevated ground. The equipment used in the fieldwork was DDR 3 manufactured by IGIS Ltd., Hyderabad. Hitherto no attempt was made to assess its groundwater potential. The results show a well distinctive electrical resistivity signature in and around the fault zone compared to areas away from study area. Based on these studies, a potential groundwater zone with a saturated thickness of about 30 to 40m and resistivity ranging from 30 to 50ohm-m was delineated. The width of the fault zone was found to be 150m. Aquifer transmissivity obtained through pumping tests on the wells a wide variation ranging from 204 sq m/day for those located on the fault zone to 80 sq m/day for the areas away from the fault zone, which shows that in such terrains the groundwater occurrence is controlled by geological structures rather than lithology.

Key words: Geological structures, Cuddapah basin, Dipole-dipole, Three electrodes

Introduction

The Cuddapah basin with its crescent shape occupies a large area of 34,560 sq km in the southern part of Andhra Pradesh State and is presumed to be a continental basin formed during upper protcrozoic period. The basin comprises of older Cuddapah and younger Kurnool formations. Many researchers have done commendable work to understand the structure and sedimentation of the Cuddapah basin notable among them are Narayana Swami (1966), Vijayam (1968). Balakrishna (1974) and Kamal (1981). The eastern boundary of this basin is severely faulted, whereas the western margin is comparatively less disturbed. However, the dislocations and diastrophic movement experienced in the eastern part of the basin has their repercussions on the western margin also and the Gani-Kalva-Veldurthi fault is one such resultant feature.

Geology

The present study area is on the northwestern part of Cuddapah basin and is around Veldurthi [longitudes E 77°57' and latitudes N15 °34'] (Fig.1), Kurnool district, A.P. Kurnool system of rocks overlies the Cuddapah system of rocks, which are represented by Gulcheru quartzites and Vempalle limestones. The stratigraphic succession of the area is given in Table 1. The study area is covered by older Cuddapah formations and to some extent by younger Kurnool formations. Gani-Kalva-Veldurthi is a major transverse fault (Balakrishna & Ramana Rao, 1974) which runs for about 60 km between Veldurthi and Chennakkapalli in a roughly E-W direction.

System	Series	Stage
	Kundair	Koilakuntla limestones
	Panyam	Pinnacled quartzites
Kurnool	Plateau quartzites	
	Auk shales	
	Jammalamadugu	Narji limestones
	Banaganapally	Banaganapally sandstones
Cuddapah	Papaghni	Vempalle limestones & shales
	Gulcheru quartzites	

Resistivity surveys

Geoelectrical investigations using detailed dipole-dipole and three electrode array profilings and Schlumberger vertical electrical sounding techniques were conducted for delineating the lateral extent of the fault zone and to decipher the productive zones, if any for groundwater exploration.

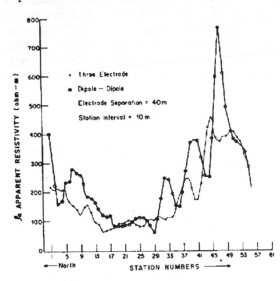

Figure 1: Apparent resistivity profile at Veldurthi over a falut

In the dipole-dipole and three electrode array profilings, the electrode separation is kept constant and the electrode array is moved as a whole with the centre of the configuration occupying successive points along a traverse. Thus, the lateral variations of resistivities of the ground can be measured.

The three-electrode array has one of the largest depth of investigations – about twice that of Wenner and Schlumberger (Apparao & Roy, 1973). In three electrode array, one infinite electrode (i.e., one current electrode) is kept at a perpendicular distance from the centre of the resistivity profile and at far off distance, usually more than 10 times to the effective electrode (i.e., one current electrode) separation from the centre of profiling.

Two profiles, one with dipole-dipole electrode array and the other with three-electrode array were carried out, traversing a length of 550m from north to south across the fault. An electrode spacing of 40m along with a station interval of 10m was maintained. The results of profiling on the fault with dipole and three electrode arrays are plotted in Fig.1. As the traverses are made from north to south (corresponding from left to right in Fig.1), apparent resistivity value is about 60 to 270 ohm-m on the north and towards south the apparent resistivity suddenly increases to higher values of more than 750 ohm-m. A very distinctive conductive anomaly is noted between the stations 15 to 29 (Fig.1). The sharp rise in apparent resistivity values on southern side of the traverse is inferred due to the presence of the Gulcheru quartzites. The sharp variations in high resistivity zone is due to the presence of semi-fractured / fractured Gulcheru quartzites. VES was carried out near failed bore well and the resistivity values correlated with lithologs of the failed bore well, which has fallen on the profile.

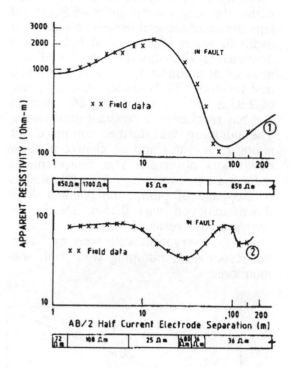

Figure 2: Geoelectrical sounding curves and interpreted layer parameters

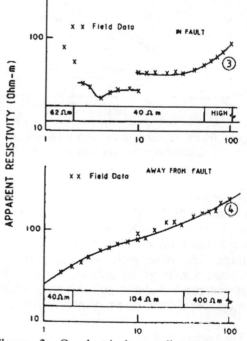

Figure 3: Geoelectrical sounding curves and interpreted layer parameters

It is observed that for the same electrode spacing, the dipole-dipole anomaly becomes less, which the three-electrode

resistivity low becomes wider. This can be explained by the fact that for thick conductive formations with vertical contact, the width of the dipole-dipole and three electrode anomaly is nearly equal to the actual width of the conducting material (target), while for the three electrode response, it is too wide by an amount approximately equal to the spacing of electrodes. Dipole-dipole electrode array and three electrode array profiles indicate high apparent resistivity values away from the fault zone to low apparent resistivity values over the fault zone. From an analysis of the profiling data, the width of the fault zone is estimated to be around 150m.

Five Schlumberger vertical electrical soundings (VES) right on the fault zone and other five soundings on either side of the fault zone were conducted in order to find out the vertical variation of resistivity with depth. The VES curves were interpreted using the auxiliary and curve matching techniques (Orellana & Mooney, 1966). The interpreted layer parameters for some VES, which are, located both within and outside fault zone are shown in Figs.2, 3, 4 & 5. It is observed that on the fault zone the VES curves are not in proper shape as the area is disturbed due to faulting and the resistivities vary from 25 to 50 ohm-m up to a depth of about 50m, but for those away from the fault zone, VES show the regular type of shapes which could be matched with the theoretical type curves. The resistivity varies from about 150 ohm-m up to a depth of around 40m. Based on the results of profiling and VES, recommendations to drill bore wells within the 150m width fault zone were made.

Groundwater occurrence

Geomorphologically the quartzites and shales of the Cuddapah system have dip amount of about 20° roughly towards east, allowing larger run off. The Cuddapahs near the study area represented by quartzites and limestones have steep dips up to 30° towards north. The region around this fault has not been explored for groundwater as it occupies an elevated ground surface. The disposition of the beds around the study area is also not conductive for the groundwater movement and storage. The total depth of wells near Veldurthi village range from 30 to 40m and the static water levels are around 15m below ground level (bgl) with yields ranging from 500 to 1000 gph.

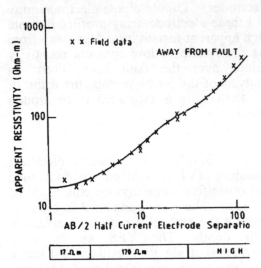

Figure 4: Geoelectrical sounding curves and interpreted layer parameters

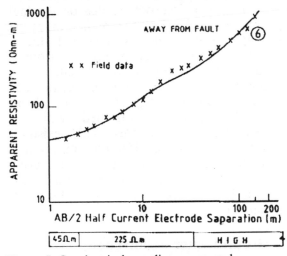

Figure 5: Geoelectrical sounding curves and interpreted layer parameters

To ascertain the control of the fault over the groundwater movement, pumping tests were conducted on the bore wells located in and away from the fault zone. The details of the tests are described below.

The total depth of the bore well located within the fault zone is about 46.5m and it taps the quartzite and limestone only. The depth to water level is 12m bgl and the diameter is 15.54cm. The bore well was pumped at a constant discharge of 400 cu m/d for about 320 minutes. A draw down of 2.93m was created during 248 minutes and has reached to a constant draw down. The pumping was further continued for about 75 minutes and no change in draw down was observed. The measurements indicated a fast recovery. After 80 minutes of stoppage of pump, the residual draw down achieved was 0.25m. Data were analysed by residual recovery method. Since no observation wells were available, the response of pumping well itself was monitored.

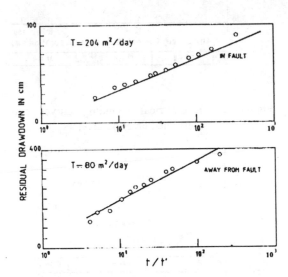

Figure 6: t/t' residual draw down for the well located within and outside of the fault zone

A second test was conducted on a bore well situated at a distance of 200m from the Veldurthi well, which is located outside the fault zone on way to Veldurthi village. The bore well was pumped at a constant discharge of 200 cu m/day. The test was run for about 200 minutes, which has created a draw down of nearly 5m.

The recovery data of both the wells were analysed by using the program – SATEM developed by Boonstra (1989). The transmissivity was obtained to be 204 sq m/d for the fault zone well and 80 sq m/d for the well away from the fault zone. The interpreted data is shown in Fig. 6.

Conclusion

Dipole-dipole electrode array and three-electrode array profiling techniques are quite useful in delineating the width of fault zones. The saturated fractures indicate a low resistivity. Dipole-dipole electrode array has better resolution in identifying the width of the fault zone.

The VES curves are not in proper shape within the fault zone and cannot be interpreted by curve matching techniques. The area within the fault zone has low resistivity and a thickness of 50m. The width of the fault zone has been calculated to be around 150m.

Acknowledgements

Author is grateful to Dr. V.P. Dimri, Director, NGRI for according permission to publish this paper and Dr. M. Thangarajan useful discussions. Dr. K. Subrahmanayam is thanked for assistance in the interpretation of pumping test data. Sri G.R. Babu, Mrs. K. Radha, Sri P.T. Varghese and Sri Joseph Gabriel are thanked for their general assistance in the preparation of the manuscript.

References

Apparao, A. and Roy, A., 1973. Field results for direct-current resistivity profiling with two-electrode array. Geoexploration, Vol. II, pp.21-44.

Balakrishna, S. and Ramana Rao, A.V., 1974. Geology and structure of Veldurthi-Gani area, Kurnool district, Andhra Pradesh. Geophysical Res. Bull., Vol. 12, No. 4

Balakrishna, S. and Ramana Rao, A.V., 1976. Vertical magnetic intensity studies of Veldurthi fault structures. Geophysical Res. Bull., Vol. 14, No. 1.

Boonstra, J., 1989. SATEM : Selected Aquifer Test Evaluation Methods - A microcomputer program. Publication No. 48, International Institute for Land Reclamation and Improvement, The Netherlands.

Kamal, M.Y. and Vijayam, B.E., 1981. Sedimentary tectonics of the Kurnool Cuddapah basin, Institute of Indian Peninsular Geology, pp. 75-85.

Narayana Swami, S., 1966. Tectonics of Cuddapah basin. Jour. of Geol. Soc. of India, Vol. 7, pp. 33-50.

Orellana, E. and Mooney, M.M., 1966. Master tables and curves for vertical electrical soundings over layered structures :Madird Intersciencia.

Vijayam, B.E. and Kamal, M.Y., 1968. Tectono-environment of deposition of the Kurnool system of Kalva-Gani area, Kurnool district, A.P. Jour. Ind. Geoscience Assn., Vol. 8, pp. 113-114

Intl. Conf. on Sustainable Development and Management of Groundwater Resources
in Semi-Arid Region with Special Reference to Hard Rock, (IGC 2002),
M. Thangarajan, S.N. Rai & V.S. Singh (Eds.)

Characterizations of aquifer systems of Kodaganar river basin, Dindigul district, Tamilnadu

V.S. Singh and M Thangarajan

National Geophysical Research Institute, Hyderabad-500 007 A.P., India

Abstract

Kodaganar River Basin lies in the Dindigul District of Tamilnadu, which is drained by river Kodaganar. Granites and gneisses essentially occupy the drainage basin. The entire area is drought prone. Shallow aquifers are continuously being exploited for industrial and irrigational usages over a long period, which has caused continuous decline in water table. There has not been proper treatment and disposal of domestic sewage and industrial effluents, which has resulted into deterioration in groundwater quality.

In order to have an assessment and proper management of groundwater resources from these shallow aquifers to meet the increasing demand, a mathematical model of flow and transport is needed which in turn requires the characteristic parameters of the aquifer system. The pumping test was carried out on large diameter irrigation wells which are fitted with pump and offers cost effective means to conduct the test. In Kodaganar River Basin, 28 pumping tests have been carried out. The details of the tests and analysis are presented in the paper. The T values are in general found to vary from 4 to 1166 m^2/d and S values from 0.00001 to 0.099.

Key words: Hard rock, aquifer parameters, pumping test, and well storage.

Introduction

The estimation of aquifer parameters namely transmissivity (T) and storativity (S) are essential for the assessment of groundwater flow regime and its potential. These parameters are also vital for the management of the groundwater resources through the use of groundwater flow model. These parameters are estimated either by means of in-situ test or performing test on aquifer samples brought in the laboratory. The applicability of the result from the laboratory test has limitations while in-situ tests give representative aquifer parameters. The most common in-situ test is *pumping test* performed on wells, which involves the measurement of the fall and rise of water level with respect to time. The change in water level (*drawdown/recovery*) is caused due to pumping of water from the well. The change in water level with time is then interpreted to arrive at aquifer parameters. Theis (1935) was first to propose a method to evaluate the aquifer parameters from the pumping test on a bore well in confined aquifer. Since then, several methods have been developed to analyze the pumping

test data (time-drawdown) under different conditions. The availability of farmer's well, equipped with pump, makes the pumping test most cost-effective. Such wells are found in abundance in most of the areas and can be used for performing pumping test.

In Kodaganar river basin, 28 existing wells fitted with pumps have been selected to perform pumping test. The pumping test data (both pumping and recovery phase) have been interpreted considering the field conditions, to evaluate aquifer parameters. The T values are in general found to vary from 4 to 1166 m^2/d and S values from 0.00001 to 0.099.

Study area

Kodaganar river basin, covering about 2000 sq km area mostly lies largely in Dindigul District and a small part in Karur District of Tamilnadu State (Fig. 1). The area is hard rock terrain and there is indiscriminate exploitation of groundwater through wells particularly for irrigation. The number of energized wells in Dindigul

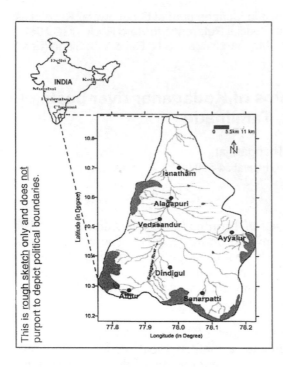

Figure 1: Location map of study area

district alone has gone up from 60 thousand in 1990 to about 74 thousand in 1998. Similarly the irrigated area from these wells has also gone up from about 60 thousand hector in 1991-92 to about 96 thousand during 1996-97, (PWD, 2000).

These figures show clearly that the demand for groundwater has consistently been increasing which has resulted into depletion of water level and thus the reduction in the groundwater potential. The other effect is deterioration in the groundwater quality. In order to assess the available groundwater potential in the area and to develop a sustainable groundwater management scheme through mathematical modeling the knowledge of aquifer parameters is vital. Therefore, estimation of aquifer parameters has been carried out through pumping tests on existing open wells.

Hydrogeological setting

The study area is drained by river Kodaganar which flows northward and joins river Amaravati (Fig.1). Most of the tributaries of Kodaganar River originate from the hills, which enclose the basin from three sides namely, east, south and west. In the southern most side, the Sirumalai Hills form drainage boundary. There is a large surface water reservoir at Attur at the foothill of the hill range. Therefore, the Kodaganar river basin is a closed river basin almost enclosed by structural hills and the entire runoff drains towards its confluence with river Amaravati.

There are two surface reservoirs one at Attur in the southern corner of up-stream and another at Alagapuri, at down-stream side north of Vedasandur. Most of the rainfall in the area occurs during the months of September to December. The average annual rainfall at Dindigul is about 828 mm and at Vedasandur about 720 mm (PWD, 2000).

Hydrogeology of the Dindigul District is described in detail by Chakrapani and Manickyan (1988) and that of Vedsandur by Balasubramanian (1980). Photo-geological study of the Kodaganar river basin has been carried out in detail at 1:50000 scale by TWAD Board (Chethia Gounder, 2000). Some of salient features are briefly described here.

Most of the study area is occupied by granite and gneisses except at southern end where charnockite hills form the drainage boundary. The larger part of the basin is occupied by the metamorphic crystalline rocks, which are highly folded, fractured and jointed. Quartzite and pyroxenite occurs in patches.

The denudational terrain surrounded by structural hills as described above, occur in the form of pediments. Shallow pediments and pediments are major geomorphic units found in the basin. These are also characterized by deep buried pediments at places. The thickness and intensity of this landform varies depending upon the slope and structural disturbances of the area. The area covered by pediment (mostly in northern part) exhibits rock outcrop with or without soil cover. These areas are basically runoff zones and the groundwater potentials in these areas are described very poor. In the shallow pediment area the groundwater potential is described as moderate. The areas of low relief constituting the buried pediment are most favorable zones for good groundwater potential. Along the sides of the river or tributaries, flood plains of recent origin are formed. A limited extent of valley fills is also found in the basin. The lineaments are found in the entire area but of limited extent.

The weathering has intensively taken place in the formation at places. The gneisses are highly weathered at places forming thick weathered zone (15 to 20 m) whereas at places the highly fractured gneisses form shallow aquifer zone. Therefore aquifer is formed of highly weathered as well as fractured gneisses. It is also evident that at places the recent alluvium along the river has formed potential aquifer.

Assessment of groundwater potential in the basin requires the knowledge of aquifer parameters and therefore, aquifer tests have been performed in the basin.

Aquifer tests

In order to assess the aquifer parameters, namely T and S, pumping tests have been carried out in the basin. The large diameter wells are found in abundance and most of these are fitted with pumps. Therefore, they offer most economic way to conduct pumping tests on them. As the well is pumped the groundwater flow towards the well becomes radially symmetrical which is shown in Fig. 2. The groundwater flow can be described as

$$\frac{\partial^2 s}{\partial r^2} + \frac{1}{r}\frac{\partial s}{\partial r} = \frac{S}{T}\frac{\partial s}{\partial t} \tag{1}$$

where, s is drawdown at radial distance r and time t. The above equation is unsteady state groundwater flow equation in homogeneous, isotropic and confined aquifer. The boundary conditions can be described as follows :

- Initial drawdown in the well is zero,

$$s(0) = 0 \tag{2}$$

- Initial drawdown in the aquifer is zero,

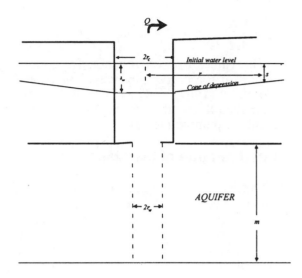

Figure 2: Flow towards the pumping well

$$s(r, 0) = 0 \tag{3}$$

- At any time the drawdown in the aquifer at the face of the well is equal to that in the well,

$$s(r_w, t) = s_w(t) \tag{4}$$

- At large distance the drawdown is zero at time t

$$s(\infty, t) = 0 \tag{5}$$

- The rate of discharge of the well is equal to the sum of the rate of flow of water into the well and the rate of decrease in the volume of water within well

$$2\pi r_w T \frac{\partial s(r_w, t)}{\partial r} - r_c^2 \pi \frac{\partial s_w(t)}{\partial t} = -Q \qquad t > 0 \tag{6}$$

where

s_w is drawdown in the well at time t
r_w is the effective radius of well screen
r_c is the radius of well casing
Q is constant discharge during the test

Using Laplace transform with respect to time and using the above conditions the solution to equation (1) can be written as

$$s = \frac{Q}{4\pi T} F(u, \alpha, \beta)$$

where, $F(\ )$ is well function.

61

$$u = \frac{r^2 S}{4Tt}, \quad \alpha = \frac{r_w^2 S}{r_c^2} \quad \text{and} \quad \beta = \frac{r}{r_w}$$

The values of well function for various values of u are given by Papadopulos and Cooper (1967). They have further described the method to calculate T and S from the pumping test data.

Interpretation of test data

During the pumping test in case of higher permeability of the aquifer the contribution from the aquifer becomes significant and well storage effect becomes insignificant. In such cases the method described by Theis (1935) can as well be used to estimate aquifer parameter. However, in the hard rock aquifers due to poor permeability and high discharge rate, for most of the duration of the pumping phase, the well storage significantly affects the total discharge rate and the aquifer discharge to the well varies with time during pumping phase as found in a case study (Singh, 2000) which is shown in figure 3.

Therefore, the variable discharge rate test cannot be utilized to interpret using Theis (1935) method, which gives ambiguous results. Further, in hard rock terrain where aquifers are of poor permeability, the aquifer response during the pumping field is almost negligible and hence it was suggested to include the recovery phase data to evaluate the aquifer parameters (Singh and Gupta, 1986). Also, in the hard rock most of the shallow aquifers are of small saturated thickness and during the pumping test there may be significant variation in the saturated thickness of the aquifer, particularly in the vicinity of the pumping well.

In order to consider the above described boundary conditions, finite difference method (Rushton and Redshaw, 1979) has been considered to interpret the pumping test data.

The method involves solving the groundwater flow equation (1) using finite difference method. The method can also be employed to take into account variety of other boundary conditions, which are common in the field. The method requires discritization of the aquifer and the test duration. The radial distance from the center of the pumping well is divided into increasingly discrete intervals *(a=logr)* as shown in figure 4. Similarly, the during of test is discritized as shown in figure 5.

The boundary condition at the well (the discharge) and at the boundary is also prescribed in the similar terms as is expressed by Equation 2 to 6. Thus finite difference expression is written

$$\frac{mk_r}{\Delta a^2}(s_{n+1} - 2s_n + s_{n+1})_{t+\Delta t}$$

$$= S\frac{r_n^2}{\Delta t}(s_{n,t+\Delta t} - s_{n,t}) + Q_{t+\frac{1\Delta t}{2}}r^2$$

where s_n is the drawdown at the nth node of radial distance r and time t, k_r is the hydraulic conductivity and m is the saturated thickness of the aquifer.

The above equation when written at various notes of the model forms as simultaneous equations, which may be solved for drawdown.

The well storage is considered by assuming that aquifer extends into the region of the well. The properties of this region are considered differently so that it represents free water into the well in this model the horizontal hydraulic resistance ($\frac{\Delta a^2}{mk_r}$) and time resistance ($\frac{\Delta t}{Sr_n^2}$) at the note representing well area, are suitably modified to represent free water in the well. Initial guess values of aquifer parameters are used to calculate the time drawdown/recovery and matched with the observed time drawdown/recovery. The aquifer parameters are then varied to get a close match between observed and calculated time drawdown/recovery. The best fit of these curves gives representative aquifer parameters.

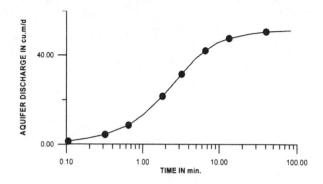

Figure 3: Variation in aquifer discharge rate

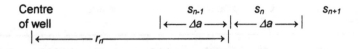

Figure 4: Radial finite difference mesh, logarithmic increase in mesh

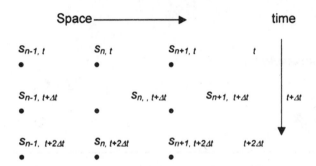

Figure 5: Representation of discrete time

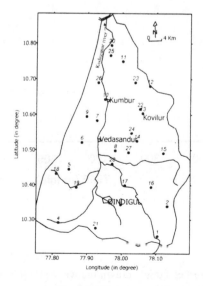

Figure 6: Location map of pumping test sites

63

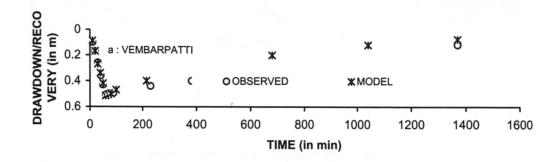

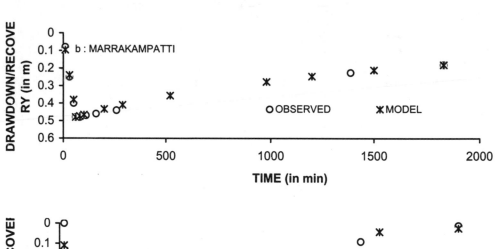

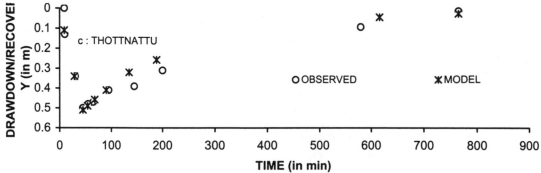

Figure 7a,b and c: Time – drawn/recovery plot (observed and model)

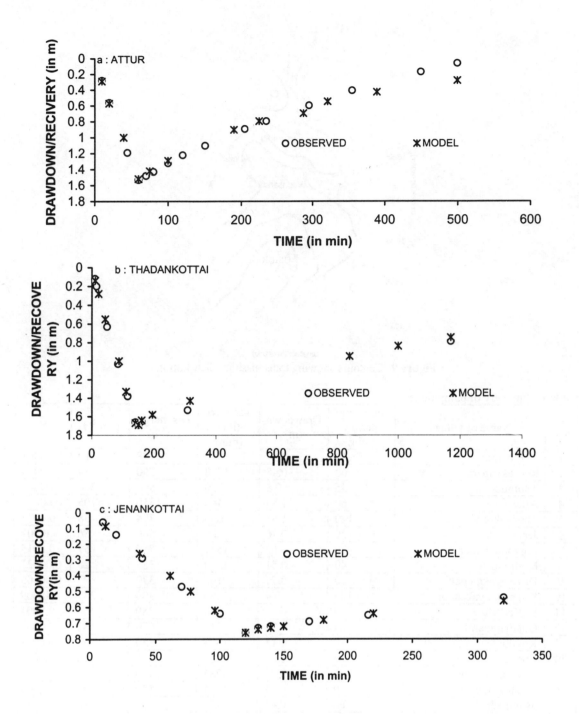

Figure 8a, b and c: Time-drawn/recovery plot (observed and modeled)

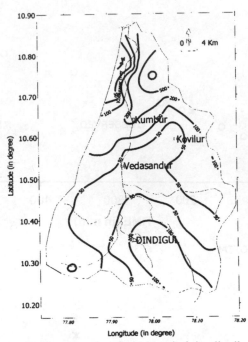

Figure 9: Contours showing transmissivity distribution

Table 1: Summery of pumping test

Sl. No.	Name of village	Pumping period (min)	Drawdown (m)	Recovery time (min)	Discharge (m³/d)	T (m²/d)	S
1	Vembarpatti	60	0.51	1310	778	110	0.00001
2	Marrakkapatti	60	0.48	1770	750	39	0.0004
3	Thottumattu	45	0.5	720	590	200	0.00001
4	Attur	60	1.53	440	1350	105	0.00075
5	Thadankottai	140	1.66	1030	794	38	0.0035
6	Jenankottai	120	0.76	200	560	56	0.009
7	Rajgoundanoor	30	0.62	1541	730	57	0.0015
8	Mathinipatti	40	0.51	1570	387	25	0.00021
9	Devinayakampatti	30	0.71	100	370	187	0.00305
10	Kirappanayakanoor	120	0.4	1350	900	237	0.009
11	R. Vellodu	90	0.32	70	780	1166	0.0001
12	Gujaliamparai	25	0.26	310	620	182	0.0001
13	Kovilur	60	1.18	1338	550	32	0.0001
14	Eriodu	60	0.76	350	420	70	0.0005
15	Pallakkurichchi	60	0.73	285	950	84	0.00004
16	Singarkottai	60	0.71	1314	540	25	0.00001
17	Mullipadi	60	0.41	170	500	96	0.00031
18	Srirampur	121	1.23	1671	978	55	0.0005
19	Mangarai	145	1.7	1400	673	4	0.0005
20	Kurumbapatti	120	1.28	330	825	118	0.0035
21	Ambathurai	120	1.16	1180	414	15	0.00035
22	R. Pudukottai	120	1.93	320	776	25	0.0057
23	Karipatti	120	1.71	300	676	119	0.00007
24	Ammapatti	80	0.93	1435	655	8	0.007
25	Kuthanayakanputhur	120	2.22	1335	2155	108	0.0055
26	Andipatti	240	1.13	315	525	56	0.099
27	Arupampatti	45	0.73	2825	660	5	0.0031
28	Undarpatti	135	0.9	305	492	53	0.0095

66

Pumping tests on wells

Most of the existing wells are fitted with pumps in Kodaganar river basin. Although, these wells are pumped regularly for irrigation purposes, some of these are selected such that owners can stop pumping for couple of days prior to test so that the sufficient recovery takes place in the well. Only such wells are selected which do not have nearby pumping well and discharge from the well is diverted far away from the well. It has also been kept in mind that the inner diameter of the well remains regular and same over the length for which water level changes. The summery of the tests are given in table 1.

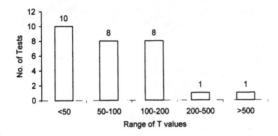

Figure 10: Frequency distribution of T-values

Discharge from the test well has been monitored at regular interval and it has been kept constant. A known volume container has been used to measure the discharge rate with the help of stop watch. Discharge rate during each test has been found to be constant whereas it has been found to vary from 370 to 2155 m³/d from test to test. In most of the cases the recovery was attempted to be observed for as much longer as possible, however the need for again pumping the well by the farmer for their crop made us to discontinue the recovery observation. The recovery time varies from 70 to 2825 m in these tests. Location of these test sites is shown in figure 6.

Results

Numerical method was used to interpret pumping test data considering both pumping as well as recovery phase. The initial guess values of T and S were used to calculate time drawdown/recovery data, which were matched with the observed time drawdown/recovery. The values of T and S were then progressively changed to get a close match between observed and calculated time drawdown/recovery. The best match between observed and calculated time drawdown/recovery has resulted into representative aquifer parameters which has given in Table 1. The plot of best fit between observed and calculated drawdown/recovery are shown in figure 7 and 8. The contour values of transmissivity for different ranges are shown in figure 9.

Out of total 28 tests, 10 tests (36%) show T value less than 50 m²/d, 8 tests (28 %) each between 50 to 100 and 100 to 200 m²/d, while one (4%) each for the range of 200 to 500 and more than 500 m²/d. Therefore, in general the transmissivity values have been found to vary from less than 50 to 200 m²/d.

Acknowledgements

Rajiv Gandhi National Drinking Water Mission is thanked for funding the project "Assessment of Groundwater Potential in Kodaganar River Basin, Dindigul District, Tamilnadu". This paper is part of the project work. Authors are grateful to Director, NGRI for his kind permission to publish this paper. During the field investigations Mr. Ramesh and Mr. Radhakrishna from Peace Trust, Dindigul, were of grate help. Authors thank to them. The help render during preparation of manuscript by Mr. P.T.Verghese is acknowledged.

References

Balasubramanian, K., 1980, Geology of parts of Vedasandur Taluk, Madurai District, Tamilnadu, Progress Report for the Field season 1979-80, GSI Tech Rept, Madras, p.14

Chakrapani, R. and Manickyan, P. M., 1988, Groundwater Resources and Developmental Potential of Anna District, Tamilnadu State, CGWB Rept, Southern Region, Hyderabad, p. 49.

Papadopulos, I. S. and Cooper, H. H. Jr., 1967, Drawdown in a well of large diameter well, Water Resou. Research, Vol 3, pp.241-244.

PWD , 2000, Groundwater Perspectives, a profile of Dindigul District, Tamilnadu, Chennai, January 2000.

Rushton K.R. and Redshaw S.C., 1979, Seepage and Groundwater Flow, John Wiley Publ., pp. 339

Singh, V.S. 2000 Well storage effect during pumping test in an aquifer of low permeability, Hydrological Sciences Journal, Vol. 45, No. 4, pp.589-594

Singh V.S. and Gupta, C.P., 1986 Hydrogeological parameter estimation from pumping test on large diameter well, Journal of Hydrology, Vol 87, pp. 223-232.

Theis C.V., 1935 The relation between the lowering of the piezometric surface and the rate and duration of discharge of a well using groundwater storage, Trans. Amer. Geophy. Union, Vol. 16, pp.519- 524.

GROUNDWATER RECHARGE

Banking of stormwater, reclaimed water and potable water in aquifers

Peter Dillon

CSIRO Land and Water, PMB2 Glen Osmond SA 5064

Abstract

A water conservation technique known as Aquifer Storage and Recovery (ASR) has been used successfully in Australia to harvest waters that would otherwise be wasted to expand irrigation and industrial water supplies. In 2001, 1 million cubic metres of urban stormwater was injected into formerly brackish aquifers and reused for municipal irrigation and industrial supplies and by 2005 the installed annual capacity will be 4 million cubic metres in semi-arid areas. While these volumes are so far only a small fraction of urban water use, the concept is spreading, and based on our current research on injection of reclaimed water (treated sewage effluent), substantially larger supplies are possible. The research has focussed on the water quality changes that occur during storage of such waters in aquifers, overcoming operational problems and developing guidelines. Aquifers are not just below ground reservoirs, they serve as bioreactors and our research addresses water quality issues that are intended to lead to sustainable recovery of potable supplies while assuring protection from groundwater pollution. Advances in understanding of the fate of pathogens, organics, nutrients and inorganic substances in aquifers are being made. However, if urban aquifers are to be used as part of water supply and treatment infrastructure, modifications to groundwater management strategies in most cities of Asia, for fresh, brackish and already-polluted aquifers will be needed and the necessary policy objectives will be briefly explored.

Keywords: Groundwater, water reclamation, artificial recharge, conjunctive use, water resources development, aquifer replenishment, aquifer restoration, stormwater reuse, water treatment, water conservation

Introduction

Many cities and agricultural areas rely on conjunctive use of surface water and groundwater. However for some cities which depend solely on surface water, there comes a time when the most economic next source of supply, taking into account the environmental cost of increased diversions or new dams, is groundwater. Where the groundwater quality is unsuitable for supply, recharge enhancement in times of excess surface water, can produce new groundwater supplies of suitable quality.

Two developments are occurring in Australia which will enable more flexibility in future water supply. (1) Wastewater treatment processes are improving, and in some locations, such as Adelaide, reclaimed effluent is now more economic than developing new resources for some classes of use, such as irrigation.

(2) It is now widely realised that water treatment costs in municipal systems are based on requirements of only a small fraction of the use of reticulated water, so that dual reticulation systems, especially at smaller than traditional scales of water supply and effluent collection and treatment systems, may be commercially attractive.

This flexibility presents opportunities for more holistic urban water management, recycling more water and reducing water imports and discharges of polluted water. These concepts, actively promoted by Clark and Desmier (1991) and Argue (eg Bekele and Argue, 1994; Argue 1997) as the "new thinking in water management', and explored within the CSIRO Urban Water Program, are gradually being implemented in Australia. CSIRO, in sympathy with the Council of Australian Governments (COAG) water reform agenda (Thomas *et al*, 1997) and in partnership with members of the Water Services Association of Australia, is continuing to provide research support for implementation of full scale experimental urban water management systems. Quantifying actual costs, risks, operating requirements, public acceptance, and financial and environmental benefits through such trials, will encourage appropriate innovation by infrastructure investors and their advisors.

COAG water reform policies also encourage competition, enhancing opportunities for small scale investors to profit from provision of alternative water supplies, while generating environmental benefits. Water banking with stormwater, reclaimed water, and potable water are vehicles, among others, to realise system flexibility, expansion of existing capacity, and increase competition, within urban water systems.

Water Banking

Water banking is a very simple concept. Just like a dam, but below ground, we store surface water during times of excess and recover it in times of shortage. Storage in aquifers can occur by infiltration from ponds and channels, or, under some conditions, by injection via wells, as demonstrated for potable supplies by Pyne(1995). Figure 1 represents injection of reclaimed water into saline aquifers and subsequent recovery for irrigation supplies. Where the same well is used for injection and recovery, Pyne termed this aquifer storage and recovery (ASR). Separate injection and recovery wells may be used where ambient groundwater quality supports the beneficial uses for which the water is recovered. This gives the benefit of additional water treatment within the aquifer between injection and recovery.

ASR with stormwater, reclaimed water and potable water

A review of international experience in ASR (Pavelic and Dillon, 1997) identified 45 case studies, including 70 known sites in 12 countries, with published information on; site characteristics and

recharge techniques; operational problems such as clogging, and means to resolve these; and monitoring of impacts on groundwater quality or the quality of recovered water. Of the 45 case studies, 71% used "natural" source waters (rivers, lakes and groundwater), and the remainder used treated sewage effluent (20%) or urban stormwater runoff (9%) (Figure 2). Highly treated sewage effluent is commonly used in the United States of America. This yields a very consistent, high quality (but expensive) water at a relatively uniform flow rate, making this attractive as a source of water for ASR (National Research Council, 1994). Retention in an aquifer provides the necessary contact with the natural environment to make recovery for potable reuse acceptable to consumers. The sustained injection of urban stormwater was found in only four documented cases, three of which were in South Australia.

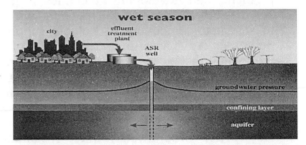

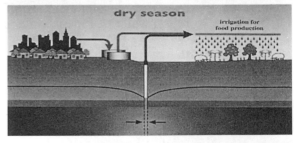

Figure 1: Aquifer storage and recovery showing injection of reclaimed water into a brackish aquifer in the wet season, and recovery of the freshened groundwater in the dry season for irrigation

The question of whether the quality of stormwater and treated effluent is adequate for ASR has been addressed and a survey of the characteristics of these classes of water, and the effects of passive treatment in wetlands reported (Pavelic and Dillon, 1995). Stormwater quantity and quality are determined by rainfall, catchment processes and human activities, which cause its flow and composition to vary in space and time. Municipal treated effluent on the other hand, is much more consistent in flow and composition, and water quality is determined by source water, the nature of industries connected to sewer, and their proportion of sewer discharge, and the effluent treatment processes. For the set of Australian samples considered stormwater had higher suspended solids, heavy metals and bacterial numbers, and lower dissolved solids, nutrients and oxygen demand than secondary treated sewage effluent. Guidelines on the quality of water for injection have been developed subsequently, as is explained later.

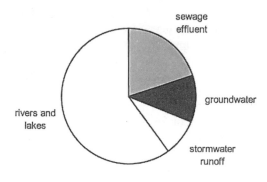

Figure 2: Types of ASR source waters *(n=45)*

Much has already been written about the benefits of ASR with stormwater (eg Dillon *et al* 1997, Gerges and Howles 1998). These have demonstrated that saline and brackish aquifers can be reshened for use as irrigation water supplies. ASR takes two un-utilised water resources, and adds value to them both by blending them at times of excess supply, storing until times of peak demand, and then recovering the water for the highest valued uses to which it can be applied. The economic viability of ASR with stormwater may depend on the establishment of urban stormwater detention ponds for flood mitigation or other purposes.

The viability of ASR with post-secondary treated effluent is currently being evaluated on the Northern Adelaide Plains (Dillon *et al*, 1999). This will make use of installed effluent treatment capacity for direct reuse of reclaimed effluent for irrigation of food crops. Effluent supply is relatively constant throughout the year, but irrigation demand varies, and effluent supply in winter exceeds irrigation demand. Storage of surplus treated effluent in depleted aquifers over winter will enable it to be recovered in summer to meet peak demand, and thereby allow expansion of the irrigated area. A distributed network of ASR wells will enable this expansion without increasing the capacity of trunk pipelines in the irrigation distribution system (Gerges, 1996).

In USA, UK, Netherlands and Israel, ASR with mains water or its equivalent is practised. This meets peak water demands when these exceed water treatment capacity, and storage within the distribution system is small. Surplus treatment capacity in the off-peak season is used to treat water for injection into an aquifer, for subsequent recovery to the distribution system at times of peak demand, usually with minimal post-treatment (eg chlorination). In one UK system, ASR is used for mains pressure compensation even on a daily basis.

An unexploited aquifer underlying or near a city is latent water resources infrastructure, which has a capacity to store, treat, and distribute water. A good aquifer, can be considered therefore as both a dam, treatment plant, and reticulation network, for which the capital cost is a comparatively trivial access and restoration charge, and the operating costs involve pre- and post- treatment, pumping, and monitoring costs. The access charge is simply the cost associated with drilling a well. Restoration costs are negligible where the groundwater is of suitable quality for the intended use of the recovered water. However in arid and semi-arid areas aquifers are commonly brackish or saline, and some injection of water is required to establish a water quality buffer zone. Depending on the type of aquifer and regional groundwater flow, this volume could range from less than the anticipated mean annual rate of injection and recovery, to a number of times that volume. In some cases an annual allocation for maintenance of the buffer zone will be needed. The cost of injecting this unrecovered water is the restoration cost. One branch of ASR research in CSIRO aims at more quantitative prediction of restoration requirements.

Improving the environmental value of the groundwater resource should be an aim of water banking. While drinking water has a higher resale value than lower classes of water such as irrigation, and it has been suggested that the utility of an aquifer would be improved most by insisting that water injected needs to be of potable standard, in reality treatment of reclaimed waters to potable standards is rarely economically viable. Furthermore, this is not precluded as a future option so long as the principle of improving the environmental value of the resource is adopted. In addition, the needs of existing environmental values of the groundwater system should be taken into account, including existing beneficial extractive uses of the groundwater and ecosystem support values. That is, the potential for negative aspects of ASR on groundwater quality and pressures should not be overlooked.

This paper advocates water banking , not disposal of unwanted water into aquifers without thought of reuse. The latter may have significant and undesirable impacts on groundwater quality and pressures. The recovery element is important for maintaining a longer term hydrologic equilibrium in the aquifer and ensuring that there is an ongoing vested interest in the quality of the injectant.

Experience with stormwater ASR

In South Australia research and development initially concentrated on injection of stormwater into limestone and fractured rock aquifers. A series of ASR sites has been established by the SA Department for Water Resources, SA (Table 1). Drainage wells in Mount Gambier have been in operation for more than 100 years, and a water quality assurance program led by the EPA aims to ensure that injected stormwater will not irreversibly contaminate the unconfined Gambier limestone aquifer which contains potable water in this area.

Previous accounts (Dillon *et al* 1997, 1998, Pavelic *et al* 1998) describe ASR operations at Andrews Farm, a northern suburb of Adelaide. These commenced in 1993 and have been intensively monitored from 4 wells over a period of five years. Water was harvested from a stormwater detention basin, and injected into a Tertiary limestone confined aquifer containing brackish groundwater which in that locality was considered too saline for irrigation. A total of 240,000m^3 stormwater was injected over four winter seasons, and the first major recovery of 100,000m^3 was undertaken in 1997-98. The results are most promising, as clogging problems have been overcome, water quality in the aquifer has been protected or improved, and the quality of recovered water is satisfactory for irrigation.

Results from Andrews Farm and the other SA sites, together with international reviews of ASR practice and regulations, and various fundamental guidelines of the National Water Quality Management Strategy were used to produce Australian guidelines for the quality of stormwater and treated effluent for injection into aquifers for storage and reuse (Dillon and Pavelic, 1996). These are different to other guidelines in use elsewhere in the world as they allow other beneficial uses besides drinking water, and they allow for the sustainable capability for treatment (notably for pathogen attenuation) within the aquifer. These guidelines were disseminated, along with other information including how to assess the suitability of prospective ASR sites at several workshops since 1996. An overview of the development of the guidelines is presented in Dillon and Pavelic (1998).

Table 1. Summary of Aquifer Injection Operations in South Australia

Site (Year of commence-ment)	Aquifer Type (Reference)	Source of Water	Infrastructure	Bore Recharge Rate L/sec	Annual Recharge Volume ML
Mt Gambier (late 1800's)	Tertiary Limestone (Telfer, 1995)	Stormwater	300 drainage wells		2800
Andrews Farm (1993)	Tertiary Limestone Confined (Pavelic & Dillon, 1996; Dillon et al., 1997; Gerges et al,1996, Pavelic et al, 1998a)	Stormwater	wetland 1 injection well 3 obs. wells	15 - 20 (gravity or pressure)	100
Greenfields (1995)	Tertiary Limestone Confined (Gerges et al., 1996)	Stormwater	wetland 1 drainage well 1 obs. Well	10 -15 (gravity)	100
The Paddocks (1995)	Tertiary Limestone Confined (Gerges et al., 1996)	Stormwater	wetlands 1 injection well	8	75
Angas Bremer (mid 1970's)	Tertiary Limestone (Gerges et al., 1996)	River flow	~30 drainage wells	5-40 (gravity)	1000
Clayton (1995)	Tertiary Limestone Confined (Gerges et al., 1996; Gerges & Howles, 1998)	Lake	1 injection well 7 obs. wells	40	70
Northfield (1993)	Fractured Rock (Stevens et al., 1995)	Stormwater	wetland 1 drainage well 1 obs. Well	10-15 (gravity)	40
Scotch College (Torrens Park) (1989)	Fractured Rock (Armstrong, pers. com.)	Creek flow	1 injection well 1 production well	15 (gravity)	40

In Bandung, Indonesia, CSIRO and local partners have commenced preliminary research on the potential for reversing groundwater declines by injection of roof runoff. There are numerous other opportunities for ASR to prevent land subsidence and saline intrusion for cities located on deltas and over-using groundwater from confined aquifers. The main limitation in Asia is the quality of the water available for injection, but by being selective with sources of water, such as in Bandung, and with the type of aquifers into which to inject, or to exclude drinking as a potential use of recovered water, it is possible to manage water banking so that substantial benefits can accrue safely and sustainably.

Experience with reclaimed water ASR

A three-stage research project commenced in July 1997 to assess the potential for ASR with post-secondary treated effluent. A description of the project and its context is given in Bosher and Kracman (1998). The first stage was to determine through laboratory experiments on cores of aquifer material, using water like that intended to be injected near the Bolivar STP, whether such injection would be technically, environmentally, and economically sustainable. It was recognised that reclaimed water differed from stormwater in its nutrient and microbiological content, and these may affect clogging (Rinck-Pfeiffer et al, 1998), transport and viability of pathogens (Pavelic et al, 1996,1998b), and biogeochemical reactions within the aquifer (Rattray et al, 1996). Furthermore the potential impacts of ASR on adjacent aquifers and on pressures and quality in wells of other groundwater users in the area needed rigorous assessment. This work is currently underway, and is described in Dillon et al (1999).

A sound understanding of the physical, chemical and microbiological processes occurring within the aquifer is important to manage ASR systems with confidence. The results from a well-instrumented and monitored injection trial at this site will be used to test models of these processes, with the intention of using these to assist in the design of reclaimed water ASR operations at other sites, using site-specific data. Toze et al (2001) describes the attenuation of pathogens, disinfection by-products and selected trace organics at ASR sites by synthesising information from laboratory experiments and field ASR sites.

Experience with potable water ASR

Early experience with potable water ASR was obtained by Miles (1952) who injected drinking water from water supply mains into aquifers via several wells in the Adelaide metropolitan area. (This followed several droughts during which groundwater was pumped into the mains to supplement supplies.) Miles' results for pressure injection were promising, but presumably due to the low frequency of reservoir spills, the costs of setting up injection wells, and the uncertainty over water quality which could be recovered after a period of storage in the aquifer, his

plan for ASR wells established on a 3km grid across the Adelaide Plains languished. The concept of restoring aquifers was not considered and opportunistic recharge of brackish or depleted aquifers using mains water in occasional wet years has not yet been implemented. This would create emergency supplies and increase groundwater reserves where they have economic value.

Run of river winter flows say from the River Torrens could also be injected into fractured rocks and alluvium on the Eastern part of the Adelaide Plains to increase summer base flows in the river, improving the health of river ecosystems, and compensating for the effects of dams upstream.

Experience in injecting raw drinking water into a highly saline (35,000 mg/L) unconfined aquifer at Clayton (Gerges and Howles, 1998) has revealed that the recovered water, although marginally more saline than the injectant, is still within National Water Quality Management Strategy guidelines, and has substantially lower turbidity than the lake water. Filtration within the aquifer is an appealing aesthetic benefit of ASR. Opportunities for water banking adjacent long pipelines are also being considered in SA as a means of augmenting water supplies to remote towns in arid areas at times of peak demand.

In wetter areas, where reservoir spill may be more frequent, potable water ASR could be more attractive than reclaimed water ASR as a water resource conservation strategy, especially if reclaimed water treatment capability is limited. Mains water ASR may also be a practical strategy to reduce land subsidence in coastal areas drawing groundwater for drinking supplies, industry, or aquaculture, such as commonly occurs in overexploited delta deposits in South East Asia. Where aquifers are used for drinking water supplies, mains water ASR is recommended in preference to the other methods, for which restrictions may be imposed on the recovery of water for drinking supplies.

To capture these opportunities and develop economic strategies will require a good working model of the groundwater system, a forecast of the amount and frequency of availability of surplus mains water, and an assessment of the economic and environmental benefits which would be produced, including reduced risks of failure, and relate these to the capital and operating costs. Models of mixing and recovery efficiency which are currently being developed, would be an important part of such economic and technical assessments.

Outcomes of Australian ASR research

The Australian research on ASR commenced in 1992 with a project to assess the potential for ASR in the upper Quaternary aquifer beneath the Adelaide metropolitan area. This was followed with a field experiment at Andrews Farm (Dillon et al, 1997). At this site, an injection well was constructed in Tertiary limestone near a stormwater detention pond, and following storm events over the next four winters, water was pumped into the well from a pump mounted on a pontoon moored in the detention pond. Groundwater was initially brackish having a salinity of more than 2,000 mg/L. The injected water was fresh (<200 mg/L TDS) but turbid, and in the winter when most water was injected (100 ML) its suspended sediment concentration averaged more than 150 mg/L. It was considered quite extraordinary that the well would accept that quality of water given that ASR operators in the Netherlands, England and USA regard source water turbidities more than 2-5 NTU as unacceptable. Furthermore, unlike convention-al practice in USA the water was not chlorinated prior to injection. Even in the newly released standard guidelines for Artificial Recharge of Groundwater (EWRI/ ASCE, 2001), it is recommended that injectant is disinfected to prevent bioclogging of the well. It became obvious through further research that there were processes occurring in the aquifer that counteracted the effects of physical clogging by

suspended solids and bioclogging by micro-organisms.

Although our initial efforts at ASR were successful, in retrospect they seem primitive in comparison with best practice that has subsequently emerged from our research. At the Andrews Farm pilot project, there was a detention storage, not a wetland. The pond sides were too steep, and our ASR operations gave a range in water levels that was too large for reeds to establish. Consequently on windy days wave action on the bare clay banks resuspended some clay. Furthermore it was found that drying of the clay liner of the pond had resulted in cracks that were infilled by coarser solids during stormwater inflows, and consequently the pond leaked severely. Calculations showed that over the four years of stormwater injection, as much water seeped through the pond bed to the unconfined aquifer as was injected into the underlying confined aquifer (240ML). Wetland designs have improved, and for the same volume of cut and fill much better passive improvement in water quality can occur, and as much active storage can be produced, even when restricting this to 100mm for maintenance of aquatic and riparian vegetation. Engineered water management improvements have also developed from just the use of a coarse screen to prevent diatoms clogging the well during blooms in the wetland in spring and autumn, to parallel rapid sand filters, and control systems that can shut down injection if turbidity or electrical conductivity lie outside specified tolerances, and trigger redevelopments to prevent well clogging. These have been incorporated in a multi-barrier approach designed to protect groundwater quality (Dillon et al, 2000) (Figure 3).

Examples of how these protections may be achieved in practice for a stormwater ASR project are shown in Figure 4. Stormwater can be sourced selectively. In Figure 4 this is shown as having a capability to divert water from a stream or drain into an off-stream wetland based on the quality (eg turbidity) of the flow. This could also occur, for example by selecting only roof catchments and not collecting runoff once it had reached the ground. Selecting an aquifer that is brackish gives much more room for flexibility on the quality of injectant than targeting an aquifer that is already of potable quality, where injectant quality would need to be consistent with drinking water uses. Maintenance of equipment, such as recalibration of sensors and checks on control systems are needed and contingency plans need to be formulated and communicated so that all know how to deal with polluted water entering the injection system or aquifer, for example.

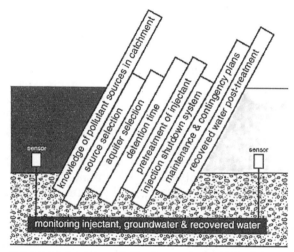

Multiple barriers to protect groundwater and recovered water at ASR projects.

Figure 3: A series of barriers has been developed to protect the quality of groundwater and recovered water at ASR sites

An important consideration that has developed from this work is that proponents of ASR projects need to ensure that proper account is taken of the monitoring and management costs of the project when evaluating the costs and benefits and deciding whether to proceed. In this way there is a clear economic incentive for the operator of viable projects to keep them well-maintained. It would be very unfortunate to have operators cut corners, as when this occurs, not only are projects likely to fail (eg via clogging), but there is also insufficient data to determine the exact cause of failure and whether and how it could be fixed. An international literature review, together with the Andrews Farm data led to Australian guidelines for the quality of

water for injection and recovery into aquifers being published (Dillon and Pavelic, 1996). These are not part of the National Water Quality Management Strategy series of documents, but rely on the principles of the Strategy, and went through review by state government natural resource and environment regulators in all states.

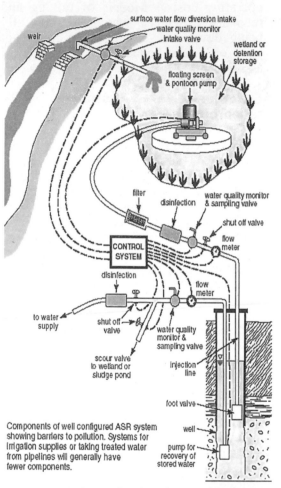

Components of well configured ASR system showing barriers to pollution. Systems for irrigation supplies or taking treated water from pipelines will generally have fewer components.

Figure 4: Schematic showing components of a well-designed stormwater ASR system

Aquifers that have been demonstrated in Australia as suitable for ASR are limestone and fractured rock. Open-hole completions are the easiest form of well completion for ASR well maintenance. However there is still insufficient research on wells in unconsolidated or unstable media requiring a screen and gravel pack to provide definitive advice on well maintenance. American experience with use of disinfected injectant in such wells indicates this can be very successful (Pyne, 1995), and Dutch experience with no disinfectant and very low turbidities and nutrient concentrations also works well.

Other outcomes of ASR research, besides the operational stormwater ASR sites and the guidelines, are maps of ASR potential for some regions and the adoption of ASR as part of state water plan for South Australia (SA Department for Water Resources, 2000). It has also resulted in the development of new techniques; to monitor pathogen survival in aquifers, to measure the ability of low-grade waters to clog aquifers, to measure hydraulic conductivity in 3D at in-situ effective stresses, and has advanced knowledge of physical and biological clogging processes (2 PhDs) geochemistry of ASR in limestone aquifers, and mixing processes and prediction of recovery efficiency in heterogeneous aquifers.

Conclusions

Water banking with stormwater has been proven to be viable and will no doubt develop as a fringe benefit of establishing stormwater detention ponds for flood mitigation in urban areas. It has wide applications for industrial estates, and municipalities with significant irrigation demands. Water banking with reclaimed water, has proved viable for highly treated effluents elsewhere in the world for recovery as drinking water supplies. The Australian studies currently underway are testing viability for irrigation quality of reclaimed water. This will have significant potential in generating irrigation resources from otherwise wasted winter/wet season effluent. There is also a niche for water banking with mains water as part of drought-proofing cities, or aquifer restoration or replenishment operations. This may be particularly attractive where groundwater is potable. Combinations of these techniques may be used in any city where there are suitable aquifers. Applications in Asia where there are significant environmental problems such as over-exploitation of groundwater leading to declining yields, land subsidence or saline intrusion may be particularly useful, as the confined aquifers may be re-pressurised directly.

Acknowledgements

The author gratefully acknowledge the support of the the Urban Water Research Association of Australia, and Hickinbotham Homes, the CSIRO Urban Water Program, and the partners of the Bolivar Reclaimed Watrer ASR Research Project (Department for Water Resources, Department of Administrative and Information Services, SA Water Corporation, United Water International Pty Ltd, with CSIRO) in undertaking research reported in this paper.

References

Argue, J.R., 1997. Stormwater management and ESD: "water-sensitive" urban design. p21-26 in Proc. Hydrological Soc. Of SA Seminar 'Stormwater in the next millennium: exportable innovations in stormwater management', Univ of SA, The Levels, Oct 97.

Bekele, G. and Argue, J.R., 1994. Stormwater management research in South Australia. Preprints Water Down Under, Adelaide,1994, IEAust NCP 94/15, Vol 3, 305-311.

Bosher, C. and, Kracman B., 1998. ASR with treated effluent on the Northern Adelaide Plains, South Australia. 3rd Intl Symposium on Artificial Recharge, Amsterdam, Sept 1998.

Centre for Groundwater Studies, 1996. Notes of Aquifer Storage and Recovery Workshop, Glenelg, 1-2 October 1996.

Clark, R.D.S. and Desmier, R.E., 1991. Towards lower cost and more environmentally sustainable urban water systems. SA Govt. EWS Lib Ref 91/19.

Dillon, P.J and Pavelic, P., 1996. Guidelines for the quality of stormwater and treated effluent for injection into aquifers for storage and reuse. Urban Water Research Assoc. of Australia, Research Report No 109.

Dillon, P., Pavelic, P., Sibenaler, X., Gerges, N., and Clark, R., 1997. Storing stormwater runoff in aquifers. Aust. Water and Wastewater Assoc. J. Water, 24 (4) 7 - 11.

Dillon, P.J and Pavelic, P., 1998. Environmental guidelines for aquifer storage and recovery: Australian experience. 3rd Intl. Symp. Artificial Recharge, Amsterdam, Sept 1998.

Dillon, P. Pavelic, P. Toze, S. Ragusa, S., Wright, M., Peter, P. Martin, R., Gerges, N., and Rinck-Pfeiffer, S., 1999. Storing recycled water in an aquifer: benefits and risks. AWWA J. Water 26(5) 21-29.

Dillon, P., Gerges, N.Z. Sibenaler, Z., Cugley, J and Reed, J., 2000. Guidelines for aquifer storage and recovery of surface water in South Australia. (Draft, Mar 2000). CGS Report 91.

EWRI/ASCE, 2001. Standard guidelines for artificial recharge of groundwater. Environmental and Water Resources Institute, American Society of Civil Engineers EWRI/ASCE 34-01.

Gerges, N.Z., 1996. Proposals for injecting effluent for aquifer storage and recovery scheme. Proc. Int. Symp. Artificial Recharge of Groundwater, Helsinki, June 1996. Nordic Hydrologic Programme NHP Report No 38, p65-73.

Gerges, N.Z., Sibenaler, X.P., and Howles, S.R., 1996. South Australian experience in aquifer storage and recovery. Proc. Int. Symp. Artificial Recharge of Groundwater, Helsinki, June 1996. Nordic Hydrologic Programme NHP Report No 38, p75-83.

Gerges, N.Z. and Howles, S.R., 1998. ASR, Hydraulic and salinity response in unconfined / confined aquifers. 3rd Intl. Symp. Artificial Recharge, Amsterdam, Sept 1998.

Miles, K.R., 1952. Geology and underground water resources of the Adelaide Plains area. SA Dept. Mines and Energy, Geol Survey Bulletin No 27.

National Research Council, (1994) Groundwater Recharge Using Waters of Impaired Quality. National Academy Press, Washington DC.

Pavelic, P. and Dillon, P.J., 1995. Will the quality of stormwater and wastewater limit injection into aquifers for storage and reuse? Proc. 2nd Intl. Symp. on Urban Stormwater Management. Melb., July 1995. IEAust. NCP 95/03, Vol 2, 441-446.

Pavelic, P., Dillon, P.J., Ragusa, S.R. and Toze, S., 1996). The fate and transport of micro-organisms introduced to groundwater through wastewater reclamation. *Centre for Groundwater Studies Report* No 69.

Pavelic, P. and Dillon, P.J., 1997. Review of international experience in injecting natural and reclaimed waters into aquifers for storage and reuse. Centre for Groundwater Studies Report No. 74.

Pavelic, P., Dillon P., Barry, K.E., Herczeg, A.L., Rattray, K.J., Hekmeijer, P., and Gerges, N.Z., 1998a. Well clogging effects determined from mass balances and hydraulic response at a stormwater ASR site. 3rd Intl. Symp. Artificial Recharge, Amsterdam, Sept 1998.

Pavelic, P., Ragusa, S., Flower, R.L., Rinck-Pfeiffer, S.M. and Dillon, P.J., 1998b. Diffusion chamber method for in situ measurement of pathogen inactivation in groundwater. *Water Research*, 32 (4), 1144-1150

Pyne, R.D.G.., 1995. Groundwater recharge and wells: a guide to aquifer storage and recovery. CRC Press, Florida.

Rattray, K.J., Herczeg, A.L., Dillon, P.J. and Pavelic, P., 1996. Geochemical processes in aquifers receiving injected stormwater. Centre for Groundwater Studies Report No 65.

Rinck-Pfeiffer, S., Ragusa, S.R., Vandervelde, T. and Dillon, P.J.., 1998. Parameters controlling biochemical clogging in laboratory columns receiving treated sewage effluent. 3rd Intl. Symp. Artificial Recharge, Amsterdam, Sept 1998.

South Australia. Department for Water Resources, 2000. State Water Plan.

Stevens, R.L., Emmett, A.J., and Howles, S.R.., 1995. Stormwater reuse at Regent Gardens residential development, Northfield, South Australia. Proc 2nd Intl. Symp. on Urban Stormwater Management, Melbourne, July 1995.

Telfer, A., 1995. 100 years of stormwater recharge : Mount Gambier, South Australia. Proc. 2nd Int. Symp. on Artificial Recharge of Groundwater, Florida, July 1994, Am. Soc. Civil Eng. New York, p 732-741.

Thomas, J.F., Gomboso, J., Oliver, J.E. and Ritchie, V.A., 1997. Wastewater reuse, stormwater management, and the national water reform agenda. Research Position Paper No 1, CSIRO Land and Water.

Toze, S., Dillon, P., Pavelic, P., Nicholson, B. and Gibert, M., 2001. Aspects of water quality improvement during aquifer storage and recovery. AWA J. Water, 28 (5) Oct 2001.

Intl. Conf. on Sustainable Development and Management of Groundwater Resources
in Semi-Arid Region with Special Reference to Hard Rock, (IGC 2002),
M. Thangarajan, S.N. Rai & V.S. Singh (Eds.)

Artificial recharge of groundwater and aquifer storage and recovery of water and wastewater: approaches for evaluation of potential water quality impacts

C. Barber
Centre for Groundwater Studies, c/o Flinders University, PO Box 2100, Adelaide,
South Australia 5001.

Abstract

Artificial recharge of groundwater has been practiced for centuries, to allow more sustainable use of groundwater particularly for irrigation supplies. In general recharge is achieved through surface infiltration into shallow, unconfined aquifers. More recently, aquifer storage and recovery (ASR) has been developed to allow direct injection into deeper aquifer systems, including those which contain poor quality groundwater (brackish to hypersaline) thereby creating freshwater storage where none existed previously. The main constraint to recharge by both infiltration and injection is pore clogging from particulates, from biofilm formation and chemical precipitation. The latter is brought about by reactions between infiltrating/injectant water and resident groundwater, and between these mixtures and mineral phases which make up the aquifer matrix. These reactions can have adverse impacts not only on pore clogging, but also on degradation of groundwater quality, for example from solubilisation of iron, or formation of dissolved sulphides. Examples of approaches used to assess possible impact on groundwater quality are presented for an arid region in NW China where an assessment was made of storage of treated sewage effluent in an intermontane aquifer system prior to re-use. This involved on-site pilot studies and modeling. Approaches involving geochemical modeling and assessment of likely impacts of injection of potable water in confined aquifers in less arid southern Australia will also be presented, highlighting the difficulties in predicting reactions where little is known of the reactivity of injectant and groundwater with aquifer mineralogy. Pilot studies will be reported from these sites and the approaches to assess impacts on groundwater quality will be compared. An approach using relatively simple hydrogeochemical modelling for prediction and interpretation of water quality trends has shown the greatest promise in these studies.

Key Words: Artificial recharge, Groundwater, Hydrogeochemical

Introduction

Artificial recharge of groundwater has been practised for centuries, particularly in arid and desert areas where natural recharge is intermittent (Pyne, 1995). The artificial augmentation of groundwater is defined as a process by which excess water is directed into the ground, either by spreading on the surface, using recharge wells or by altering natural conditions to increase infiltration to replenish an aquifer. It is a way of storing water underground in times of water surplus to meet demands in times of shortage.

Over thirty international case studies of artificial recharge using water and wastewater were reviewed by Pavelic and Dillon (1997). Generally it was reported that most studies used surface infiltration to recharge groundwater, although injection wells and ASR were becoming more common recently, and where recharge of confined or semi-confined systems was being attempted, injection and recovery wells were the only possible option.

With ASR, the injection well is used both for injection and recovery, thus overcoming what has been a major drawback with recharge using wells - that of the need for regular well development to remove clogging solids and biofilms on well screens and adjacent aquifer media (Pyne, 1995). Single well recharge and recovery also increases the likelihood of recovering injected water, which is important where there are significant differences between water of between injected water and groundwater (eg where the latter is saline, brackish or polluted). Injection wells can be developed regularly during recovery phases, and these become self-cleaning, maintaining well capacity for both injection and recovery. Additionally, the use of recharge wells overcomes the need for large areas of land needed for surface infiltration techniques of artificial recharge. An additional major advantage of using well injection is that this opens up the possibilities for confined and semi-confined aquifers to be used for storage of injected water. The combined use of wells for both injection and withdrawal of water has greatly improved the economics of ASR.

Clogging of injection wells with solids and biofilms is still a significant factor in ASR which needs to be overcome. This can be minimised by pre-treatment of effluent prior to injection. Thus partial removal of suspended solids (SS) to levels of around 10 mg/l has been suggested, although stormwater injection with SS levels averaging 30 mg/l in Australia has been shown to be sustainable over several years in carbonate aquifers.

Surface spreading of water, particularly treated effluent to augment groundwater recharge, on the other hand, takes advantage of natural water treatment as the recharge infiltrates through the vadose (unsaturated) zone before recharging groundwater. This process has been called Soil-Aquifer Treatment (SAT) or geopurification. Often, improvement in the quality of infiltrating water is the objective of a SAT system. Water or treated wastewater extracted following SAT treatment and aquifer storage is often used without further treatment for a range of non-potable uses. The latter include recreational uses (eg irrigation of golf courses, wetland and stream recharge) as well as for landscape irrigation where direct contact of water with humans is negligible. Water which is to be used for potable purposes may require some additional treatment (Pyne, 1995).

SAT using effluent with periodic infiltration and drying to avoid clogging and to promote nitrification, denitrification and other pollutant removal or retardation should be sustainable indefinitely for a wide range of non-potable uses. Primary treated sewage effluent (ie settled effluent) has been successfully used for recharging groundwater at land-spreading sites. Kanarek and Michail (1996) report on a SAT system for treated wastewater in Israel, which has operated successfully for nearly 20 years, treating flows of 270000 m^3/d of secondary-treated sewage effluent from a population of around 1.3 million. The system has successfully achieved filtration, disinfection and nitrogen removal through nitrification and denitrification during cyclic infiltration into a sand aquifer, and provided treated water for reuse in irrigation.

The desired role of SAT is stated to be removal of those constituents which pose the greatest threat to public health. Unfortunately, some processes which remove pollutants are not efficient in a natural setting, and this is the case even in customised treatment plants. Not all pollutants present in sewage are removed or degraded at the same rates or to the same extents. Natural conditions which favour attenuation of some contaminants also favour persistence of others.

The chemical and biochemical reactions which can take place in both ASR and SAT applications are many and varied, and these are difficult to predict and assess. There is a need to develop an approach to try and evaluate likely reactions which may affect the efficiency or sustainability of artificial recharge operations, as well as adversely affecting groundwater quality or some intended reuse. This paper summarises and evaluates two approaches which have been used in pilot studies of ASR in Australia and of SAT in northwest China.

Evaluation of impacts of AR/ASR on groundwater quality

The objectives of predicting geochemical interactions during artificial recharge are varied, and can be summarized as follows:

- Prediction of contaminant transport and "breakthrough"

- Reactions between mixing aqueous phases
- Reactions between aqueous phases and aquifer matrix minerals.

There are a variety of solute transport models which can be used to address contaminant transport, and given appropriate and realistic coefficients for defining partition between phases (sorption, reaction, decay, gas/liquid exchange) then contaminant behaviour during artificial recharge can be assessed using modeling, if flow processes are also reasonably well understood. The main problem with this is in calibrating models with realistic coefficients describing contaminant attenuation.

The use of physical laboratory or test-cell models of aquifer systems (column studies, sand tanks etc) have been popular for assessing contaminant behaviour, and these also have been used for assessing likely reaction between infiltrating or injected water and aquifer minerals (eg see Rinck-Pfeifer et al, 1998). Although this approach has the advantage of control over physical and chemical conditions, it is debateable whether these reflect likely conditions during artificial recharge. By nature, physical models cannot be representative of aquifer conditions at the scale of even a pilot study of artificial recharge because of (often unknown) heterogeneity of aquifer media and groundwater flow. Physical models are perhaps best suited to obtaining data under controlled conditions on reaction kinetics. These also have the advantage of visibility and accessibility, which are major drawbacks in subsurface field investigations.

Comparison of results of laboratory column studies, plume modelling and field-scale tracer testing of degradation and attenuation of monoaromatic (BTEX) hydrocarbons (Thierrin et al, 1993) suggest that problems of scaling up and representativity are significant. Results from the testing carried out during this research (as compound half-lives determined from the tracer test, from plume modelling, and from large [100mm diameter, 2m long] columns) are shown in Table 1. It is clear that determinations made in the field at, and close to plume scale are significantly different for toluene from the much smaller scale testing using the laboratory sand columns infiltrated with groundwater from the test site spiked with BTEX compounds. Other data for benzene, ethylbenzene, and xylene isomers is not precise enough from column testing where no degradation was observed over the duration of the testing (40 days). Clearly, some degradation should have been observed over this period, if degradation rates were similar to those observed at field scale. These results, although not relating to artificial recharge, highlight the difficulties in reproducing field-scale conditions at small (laboratory column) scale, particularly where conditions are dominated by microbiological activity – essentially the conditions observed in AR/ ASR and SAT systems.

Table 1: Comparison of results of tests to assess degradation of BTEX hydrocarbons in anoxic groundwater at different scales (plume scale through modeling, from in-plume injection of deuterated BTEX tracers, and in laboratory column experiments) from Thierrin et al (1993).

BTEX compound	Compound half-life in days		
	Within-plume tracer test	Plume modelling	Laboratory Columns
Benzene	>800	>800	>42
Toluene	100 ± 40	120 ± 25	0.3 ± 0.1
Ethylbenzene	nd	230 ± 30	>42
p-xylene	225 ± 75		>42
m- and p-xylene		170 ± 10	
o-xylene		125 ± 10	>42

Field-scale reactions between mixing aqueous phases, minerals, organic carbon and ion exchange phases can be assessed relatively simply using equilibrium mixing-cell models such as PHREEQC (Parkhurst et al, 1999). Thus this type of modeling approach can be used to provide a wide range of assessment of redox and pH-dependent reactions which could impact on the efficiency of an artificial recharge system. These assessments can also be used to provide quantitative evaluation of monitoring data of artificial recharge schemes.

In AR/ASR, modelling is important for assessment of mineral dissolution reactions (eg carbonate minerals) which could locally increase hydraulic conductivity but also have an adverse impact on aquifer structural integrity by dissolution of aquifer matrix and reducing the extent of mineral cementation. Likewise, mineral precipitation would decrease hydraulic conductivity and contribute towards aquifer clogging. Redox reactions in particular could give rise to both beneficial and undesirable impacts, through infiltration or injection of often oxygenated surface water into anoxic aquifers. Thus reactions between dissolved oxygen and sulphidic minerals (eg pyrite) are likely to rapidly reduce DO, lower pH and induce pH dependent reactions such as carbonate dissolution, although the low solubility of oxygen in water limits the impacts of these reactions. Progressive lowering of redox potentials where infiltrating water contains assimilable organic carbon provides conditions for denitrification, iron and managanese reduction and solubilisation, and sulphate reduction. These reactions are well known sequences described by Stumm and Morgan (1996) and their modeling is relatively straightforward. This type of mixing cell model thus provides a potentially useful tool for preliminary appraisal of aqueous-mineral reactivities during artificial recharge, of use in preliminary planning as well as providing a sound basis for evaluation of monitoring data.

The approach has drawbacks through reliance on equilibrium being attained between aqueous and mineral phases. This may not be problematic, as models such as PHREEQC now include reaction kinetics which can be used to take account of non-equilibrium conditions. Aqueous and mineral phase heterogeneity is probably still the most significant impediment to utility of this type of model.

Traditionally, little preliminary prediction of the effects of mixing and reaction between recharged water and natural groundwater is carried out, and in general these effects are only investigated by monitoring during preliminary or pilot studies. This is perhaps not surprising considering the variability in composition of water being recharged, and the natural variability in composition of groundwater,

but mostly because of natural heterogeneity in mineralogical composition of aquifer media, and of flow processes (intergranular or fracture/fissure flow where the latter bypasses intergranular pore fluids and interaction is mainly through diffusive interchange between pore water and fractures). The challenge is to find a sensible conceptual model of aquifer behaviour during artificial recharge which provides realistic appraisal of system behaviour.

A possible approach is developed here by reference to two case studies. The first of these is a feasibility study for SAT in China which used conventional laboratory and field studies in a feasibility study to assess likely system performance. Second, a preliminary field study of potable water ASR is considered, where the model PHREEQC was used to assess system performance using field monitoring data during a pilot phase injection trial. The latter approach was based on previous experience in preliminary modelling of possible ASR schemes in Australia.

Case study – Soil Aquifer Treatment in PR China

China has one of the lowest per-capita quantities of water resources, being ranked 88th in the world. Of 479 cities investigated in China in recent years, more than 300 were facing water shortages and 40 more cities had severe shortages. Recycling of water generally is very low, and often wastewater is discharged without any treatment or only partial treatment into surface water, or onto land and thence into groundwater. Untreated wastewaters are also used directly for irrigation, which has caused soil, crop and water pollution. The latter has given rise to even greater shortages of potable water. National water management departments and environmental agencies in China have formally indicated that wastewater treatment and reuse schemes need to be developed urgently, to help in reducing the water supply problems (Barber and Zhu, 1999).

Currently in China, the wastewater discharge rate is more than 100 million m^3/d. More than 80% of this is disposed directly to the environment without any treatment, giving rise to serious pollution. As a developing country, China is unable to fund construction of a large number of municipal drainage systems and wastewater treatment plants. Therefore, natural treatment systems are being developed and implemented, especially wastewater treatment in situ followed by reuse. If this can be implemented in China, the water supply and wastewater pollution situation in Northern China, especially in the Northwest District, will be significantly eased.

Given the reliance on groundwater and the presence of extensive aquifer systems in Gansu, it was considered that wastewater treatment and reuse could usefully be combined with aquifer storage of treated wastewater, for example in winter when there was no demand for reuse of treated water for irrigation. A project was consequently set up to investigate the feasibility of using SAT at the city of Jiayuguan in northern Gansu.

The overall treatment and reuse scheme proposed for the demonstration site involved the following components:

- Pre-treatment of effluent to reduce suspended solids and provide a clear effluent from which pathogenic bacteria and viruses could be removed using UV disinfection to allow immediate effluent reuse or aquifer storage depending on demand;
- Infiltration of primary-treated effluent through shallow trenches or spreading basins which are subject to cyclic infiltration and drying to avoid biofilm formation and to promote nitrogen removal;
- Soil-Aquifer Treatment (SAT) during percolation through a vadose zone to filter and disinfect effluent, precipitate and adsorb nutrients and contaminants, nitrify ammonia during drying cycles and to denitrify effluent during infiltration cycles;
- Storage of treated effluent in the saturated zone of the aquifer system prior to recovery through pumping wells;
- Reuse of treated effluent for irrigation.

For the scheme to be sustainable and environmentally acceptable, the scheme would have the following treatment targets or objectives:

- Treated wastewater entering the saturated zone after SAT would conform to irrigation water standards to protect human health and for environmental sustainability,
- Effluent within the saturated zone of the aquifer and irrigation return flows would not impact adversely on drinking water supplies.

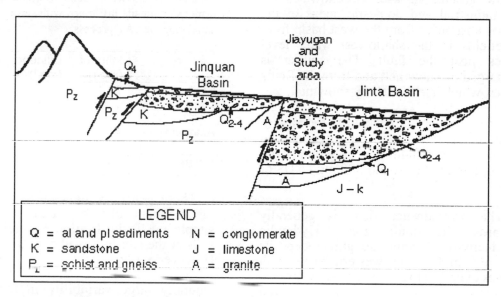

Figure 1: Schematic NE-SW geological cross-section of western and eastern basins at the Jiayuguan SAT test site, Gansu Province, PR China, showing the study area and the position of the city in the eastern basin. The vertical scale is exaggerated.

Given the difficulties in predicting quantitatively the extent of soil-aquifer treatment, and in designing infiltration and injection systems, it is always recommended that pilot schemes be operated prior to designing full-scale aquifer storage / recovery and reuse schemes. This was carried out for process optimisation at the Jiayuguan site, and biochemical reactions of the proposed scheme were evaluated through laboratory column experiments and field infiltration studies, as well as through some preliminary modelling. This is perhaps a more standardized approach which is widely applied to provide preliminary data on likely system operation, but suffers from problems of heterogeneity and scale. At the Jiayuguan test site, groundwater occurs in Quaternary and older mixed alluvial and colluvial sediments which have accumulated within a graben structure. The sediments form two basins in the Jiayuguan region. The western basin (Jinquan Basin) is separated from the eastern basin (Jinta Basin) by a fault zone and basement high on the western side of the city separating the two basins (Figure 1). In the region of the fault, the water table is close to the surface (10 to 25 m) with the thickness of the aquifer only 40 to 60 m. The aquifer is much deeper east of the fault within the eastern basin, and the hydraulic head may be found up to 100m below ground surface. Groundwater for the industrial and domestic needs of the city is abstracted from the west basin from borefields in the shallowest water level zones near the fault. The aquifer is recharged by snowmelt and from the Beida River, which itself is fed by snowmelt.

Groundwater is extracted at the eastern parts of the Jinta Basin for irrigation and domestic use. Salinities of groundwater are reported to be around 500 mg/l TDS or less.

The groundwater flow is generally towards the fault zone. Hydraulic conductivities within the gravel deposits are very high in the western basin, with values over 200 to 350 m/d in some places. Maximum yields of individual wells can be greater than 10 000 m^3/d.

The city of Jiayuguan lies within the eastern basin at its western end close to the fault. (see Figure 1). The proposed site for infiltration and treatment of wastewater is fully within the eastern basin to the southeast of the city. The piezometric head of groundwater beneath the site is around 100 m below surface, and the salinity of groundwater is less than 500 mg/l TDS. Very few wells have been drilled in this part of the basin, only four being within a 6 km radius of the treatment site, the nearest being about 3 km away.

The proposed scheme was evaluated using both laboratory and field experimental work, using sediments which were typical of those at the proposed site where infiltration was to take place. These materials were naturally quite variable and it was difficult to assess whether these indeed were representative of materials through which effluent would percolate. However, it was considered these tests would provide preliminary data on infiltration rates and hence required infiltration areas for the volumes of effluent (20 000 m^3/d), and on the possibility of obtaining the required biochemical reactions inherent in the SAT process. The results of the infiltration trials are shown in Table 2.

Table 2: Experimental results for 1.5m long laboratory columns of soils from the Jiayuguan demonstration site infiltrated with sewage effluent, simulating the SAT process

	Influent (mg/l)	Effluent (mg/l)	Removal (%)
COD	73	27	62
BOD$_5$	41-115	16-20	62-83
Ammoniacal-N	11-21	8-10	33-47
Nitrate-N	26-44	19-32	21-27

These generally indicate that a degree of treatment can be achieved through simple cyclic infiltration through natural soils at the site followed by drying cycles. However, given the expected likely infiltration area of around 15ha and naturally large variability in the glacial sediments across the proposed site, the data could only provide at best a first approximation to what might happen in reality. It was however, recognized that this work was only a part of a feasibility study. Similarly, the field and laboratory experimental work was of necessity at small scale, and given time constraints could not be extended over periods of more than a few infiltration and drying cycles. The likely representivity of those tests could also be brought into question, although it was clear that there was scope for removal of contaminants from the effluent.

Probably the main reason for abandonment of the scheme was the much greater uncertainty in predicting infiltration and storage of effluent in groundwater beneath the site. Initial information from very limited regional surveys suggested a very different hydrogeological setting to that found by preliminary drilling. Thus the sediments beneath the site were an extensive layered sequence of gravels, sands, silts and extensive clay lenses. Groundwater was possibly semi-confined, and any infiltration of effluent would have been extremely complex. With hindsight, it is clear that (at least at this site) the first appraisal should have been drilling to investigate the local conditions beneath the site. Given a favorable outcome from this, then laboratory testing and modeling and particularly a more detailed pilot investigation, at a scale more appropriate for this large-scale infiltration, would have been needed to provide data for process optimization.

Case study - evaluation of geochemical changes during potable water ASR, Jandakot, Western Australia

Conjunctive use of groundwater and surface water has been a feature of the water supply system in Perth for many years, and there are extensive unconfined (surficial aeolian sand aquifers) and confined / semi-confined Mesozoic aquifers in sedimentary rocks in the Perth Basin which are exploited for water supplies. The city is becoming increasingly reliant on groundwater, and given below-average rainfall over the last decade, is actively investigating artificial recharge options. In general, stormwater is used extensively to recharge surficial unconfined aquifers through drainage basins, to augment storage and use of shallow groundwater for irrigation of parks and gardens (Scatena and Williamson, 1999). There are perhaps limited opportunities for additional storage of potable water in the surficial sediments, as these are directly connected with wetlands and other groundwater-fed ecosystems on the coastal plain, which would be adversely impacted if groundwater levels rose significantly. However, there are deeper, confined sedimentary aquifers where potable water ASR would be feasible (Scatena and Williamson, 1999). One such scheme, and evaluation of geochemical impacts on the groundwater system and injected water, is described here.

Table 3: Simulation details for PHREEQC modelling

Injectant "storage" phase	Simulation 1	Equilibrium with phases siderite, pyrite, and ion exchange (CEC 0.03)
	Simulation 2	Equilibrium with phases siderite, pyrite, calcite and ion exchange (CEC 0.03)
	Simulation 3	Equilibrium with phases siderite, pyrite and ion exchange (CEC 0.03), and limited reaction with calcite (dissolution limited)
	Mix	No mineral phase reaction (dissolution or precitation)

PHREEQC was used to assess through geochemical modelling possible chemical reactions on mixing of groundwater and injectant (potable water from a reservoir) and reaction and exchange with possible aquifer mineral matrices during a trial potable water ASR injection into the Cretaceous/Jurassic Leederville Formation at Jandakot, near Perth in Western Australia. The study was carried out between July 2000 and January 2001, with injection of over 40 000 kl of water over 10 days into a mainly sandstone and interbedded shale aquifer. Storage of groundwater was over a 16 week period, and recovery of injected water took place over a 7 week period.

The aquifer was around 250 m thick at the site, the top of the aquifer was approximately 100 m below surface, and shale units and groundwater salinity generally increased with depth. Groundwater in the upper part of the aquifer had TDS of around 660 mg/l, whilst deeper in the aquifer, these increased to over 1900 mg/l. Both mixing and reactions during aquifer "storage" (up to week 16) and during "recovery" (post week 16) have been considered.

Conceptual models of potential mineral reactions (Table 3) were defined through consideration of mineralogical analyses of limited amounts of drillcore, from mineral-saturation relationships with natural groundwater from the site using PHREEQC, from analysis of trial injection and from over 30 initial simulations of possible water-mineral reactions. Some limited analysis of Fe(II) and Fe(III) species, sulphur species and sulphur stable isotopes (S^{34}/S^{32}) have also been used to assess speciation and reactions identified by modelling.

Reactions between mineral phases siderite, pyrite, calcite (or ankerite), and ion exchange were considered in more detailed modelling of injectant / groundwater mixes with low concentrations of organic carbon. Model output and monitoring data were compared (eg see Figures 2-4) by estimating the proportion of background groundwater to injectant based on chloride concentration, assuming this anion behaves conservatively: the fraction of groundwater (f) was given by

$$f = (Cl_{sample} - Cl_{injectant}) / (Cl_{groundwater} - Cl_{injectant})$$

Simulations were also compared with simple mixing of injectant and groundwater assuming no reaction. Sample results for dissolution, precipitation and ion-exchange by mineral phases is shown in Table 4, and actual and simulated concentrations for bicarbonate, iron and sulphate for various aqueous mixtures of injectant and groundwater are given in Figures 2-4.

Iron geochemistry and its behaviour during ASR is particularly important. Analysis and modelling indicates that Fe(II) is predominant in the aquifer, and also with injectant when equilibrated with mineral phases in the aquifer. This was confirmed by Fe(II)/Fe(III) species analysis.

The conceptual model which shows the closest fit to actual data for iron and other constituents involves dissolution of pyrite and partial dissolution of calcite (or possible ankerite) and precipitation of siderite, along with exchange of divalent cations (Ca) for monovalent cations (Na). The models also indicate that the small amounts of dissolved oxygen in injectant are utilised for oxidation of sulphide derived from pyrite dissolution, and not for oxidation of iron. These results suggest that geochemical conditions are not favourable for activity of iron oxidising bacteria.

There are some significant differences between actual and predicted concentrations of bicarbonate, iron and sulphate (Figures 2-4). Modelling indicates that sulphate reduction, which would provide additional removal of iron as iron sulphide, should be minimal due to the low concentrations of dissolved organic carbon in injectant and groundwater, and both iron and sulphate ought to increase as the proportion of groundwater increases. In fact, both decrease during the storage phase, and additionally bicarbonate increases substantially above that expected from calcite dissolution. The latter suggest that sulphate reduction and possibly iron sulphide precipitation, is occurring. This sequence of reactions is difficult to reconcile with expected thermodynamic behaviour which is modelled using PHREEQC, unless organic carbon becomes available during ASR to promote sulphate reduction following initial pyrite oxidation. Most presumed sulphate reduction takes place during the storage phase when injectant makes up the major part of the recovered water (eg 80-90%), which suggests that reduction takes place close to the injection well. It is possible that this zone may be enriched in organic

Table 4: Dissolution (-ve) and precipitation (+ve) and ion-exchange for mineral phases in simulations where f=0.1 (90% injectant) during the storage phase. Values in mMoles/L.

	Siderite	Pyrite	Calcite	Exchange phases		
	FeCO$_3$	FeS$_2$	CaCO$_3$	CaX$_2$	NaX	FeX$_2$
Groundwater	0.03	11.89	-0.23	5.65	8.65	0.017
Simulation 1	-0.292	-0.056	-	5.83	7.08	0.356
Simulation 2	0.063	-0.056	-1.53	7.17	5.38	0.013
Simulation 3	0.041	-0.056	-1	6.74	6.04	0.034

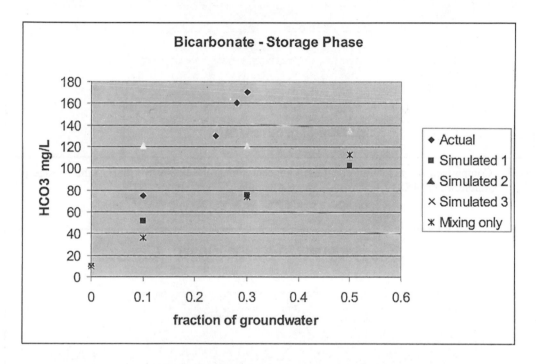

Figure 2: Comparison of actual and simulated concentrations of bicarbonate during the storage phase of potable water ASR at Jandakot, Western Australia.

matter possibly through natural filtration during injection, although this is conjectural and requires further investigation. It is concluded that the low soluble iron concentrations are at least partially due to sulphide precipitation.

Sulphate shows the greatest variability in concentration during the storage phase, and the trends over time during storage are quite complex. Modelling predicts small increases in sulphate arising from dissolution of pyrite and oxidation of dissolved sulphide by dissolved oxygen, as indicated above. However, sulphur isotope fractionation from the Jandakot site (Rattray, 2001) suggests processes of initial pyrite oxidation followed by sulphate reduction occurring during ASR, a pattern noted previously during ASR in South Australia (Rattray, 1999).

89

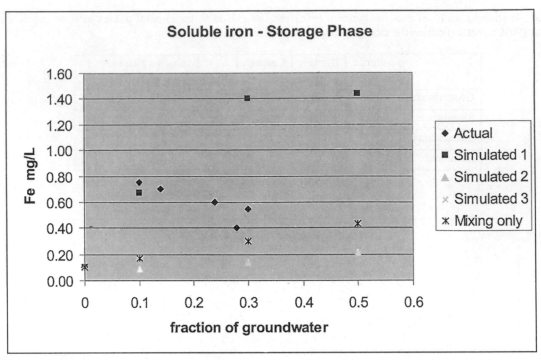

Figure 3: Comparison of actual and simulated concentrations of soluble iron during the storage phase of potable water ASR at Jandakot, Western Australia.

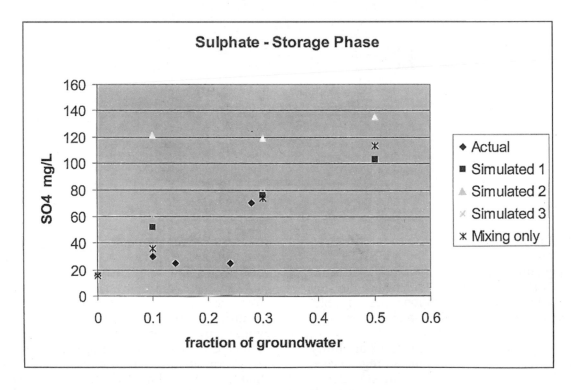

Figure 4: Comparison of actual and simulated concentrations of sulphate during the storage phase of potable water ASR at Jandakot, Western Australia.

During the recovery phase, groundwater and injectant were being pumped mainly from the upper part of the Formation, and modelling suggests lower equilibration between mineral phases and injectant and mixes of this with groundwater. Here, concentrations follow more closely the predicted trends for mixing between injectant and groundwater, with limited ion exchange, limited dissolution of pyrite and siderite and minimal dissolution of calcite. There is little evidence from modeling or S-isotope ratios that sulphate reduction takes place during the recovery phase when groundwater and injectant outside of the immediate location of the well are being recovered (Rattray, 2001). This supports the above conclusion that sulphate reducing conditions have developed very close to the injection well during the injection and storage phases.

Conclusions

It is concluded from the case study on potable water ASR, that the use of PHREEQC geochemical modelling initially in batch-reaction / mixing mode provides a useful methodology for identifying a conceptual model of water mixing and mineral reactions which impact on chemical changes of water stored in ASR schemes. However, considering the dynamic conditions close to the injection well, without sulphur stable isotope analysis the differences between predicted and actual concentrations of iron, sulphate and bicarbonate would have been difficult to interpret. It is thus clear that modeling by itself cannot be relied on to provide realistic results, but should be used as simply another tool to assist in evaluating system behaviour during artificial recharge , or as in this case during ASR.

It seems premature to suggest that deterministic modelling of water quality impacts during artificial recharge can be realistic and accurate with current understanding, largely because it is difficult to establish a conceptual model of mineral/water reactions, and because of apparent transient changes during storage very close to the ASR well. It seems likely that modeling of SAT would have similar, possibly increased difficulties compared to those experienced with the ASR case study. Similarly, physical models of SAT or ASR, even larger scale models in laboratory or field, are too small in scale to be able to provide realistic indications of system behaviour. However, these could be useful for determination of possible reaction rates and exchange capacities under controlled conditions. Even so, these may still be of questionable value, and great care is needed with use of such data in predictive modeling.

From the studies so far, the most useful approach to determination of water quality impacts during artificial recharge would involve the following:

- Identification of aqueous phase end-members and variability in space and time
- Identification of likely mineral assemblages which have importance during aquifer storage, from mineral/groundwater saturation indices for groundwater(s) in the near vicinity of the AR/ASR scheme
- Prediction of possible water quality impacts using most probable aqueous phase compositions and most likely mineral assemblages, using reaction / mixing cell models such as PHREEQC possibly in tandem with laboratory studies to assess aqueous / mineral reaction kinetics
- Use model output as simply another tool for interpretation of groundwater monitoring data, and combine these with more detailed monitoring such as determination of stable isotope ratios for sulphur, carbon, and possibly nitrogen, developing a conceptual understanding of processes and their impacts on groundwater quality in time and space.

Finally, given a good conceptual understanding of system behaviour, it may be possible to develop a deterministic model which can then be used to optimize an AR scheme, or at least to minimize adverse impacts on water quality and improve overall system sustainability

References

Barber C and Zhu K, 1999. Augmentation of water supplies for irrigation: a feasdibility study of the use of soil aquifer treatment, aquifer storage, recovery and reuse of wastewater in an intermontane basin, Gansu province, PR China. Proc. Intl Conference on Water Resource Management in Intermontane Basins, Chiang Mai, Thailand, 2-6 February 1999. Water Research Centre, Chiang Mai University.

Kanarek A and Michail M, 1996. Groundwater recharge with municipal effluent: Dan region reclamation project, Israel. Water Science and Technology, 34, 227-233.

Parkhurst, D.L., and Appelo, C.A.J., 1999, User's guide to PHREEQC (Version 2)--a computer program for speciation, batch-reaction, one-dimensional transport, and inverse geochemical calculations: U.S. Geological Survey Water-Resources Investigations Report 99-4259,312 p.

Pavelic P and Dillon P J, 1997. Review of international experience in injecting natural and reclaimed waters into aquifers for storage and reuse. Centre for Groundwater Studies report No 74, May 1997.

Pyne R D G, 1995. Groundwater recharge and wells. A guide to aquifer storage and recovery. Lewis Publishers, Florida, USA.

Rattray K J, 1998. Geochemical reactions induced in carbonate bearing aquifers through artificial recharge. Unpublished MSc Thesis, Flinders University of South Australia.

Rattray K J, Martin M, and Xu C, 2001. Aquifer Storage and Recovery at Jandakot, Perth, Western Australia. Unpublished report, Water corporation of Western Australia.

Rinck-Pfeifer S M, Ragusa S R and Vandevelde T, 1998. Column experiments to evaluate clogging and biogeochemical reactions in the vicinity of an effluent injection well. Proc. 3rd Intl Symposium on Artificial Recharge of Groundwater – TISAR 98. 21-25 Sept. 1998, Amsterdam, Netherlands, J H Peters et al (eds).

Scatena M C and Williamson D R, 1999. A potential role for artificial recharge in the Perth region: a pre-feasibility study. Centre for Groundwater Studies Report No. 84, August 1999.

Stumm W and Morgan J J, 1996. Aquatic Chemistry, 3rd Edition. Wiley Interscience, New York.

Thierrin J, Davis G B, Barber C, Patterson B M, Pribac F, Power T R and Lambert M, 1993. Natural degradation of BTEX compounds and naphthalene in a sulphate reducing groundwater environment. Hydrological Sciences, 38 (4), 309-322.

Intl. Conf. on Sustainable Development and Management of Groundwater Resources
in Semi-Arid Region with Special Reference to Hard Rock, (IGC 2002),
M. Thangarajan, S.N. Rai & V.S. Singh (Eds.)

Interconnection between water bodies in select areas of Rajasthan state, India

H. Chandrasekharan[1] and S. V. Navada[2]

1. Water Technology Centre, Indian Agricultural Research Institute, New Delhi, India
2. Isotope Division, Bhabha Atomic Research Centre, Mumbai-400 085, India
email: **chandrasekaranh@hotmail.com**

Abstract

Acknowledge of interconnection between different water bodies, caused by natural and/or man-made activities, is important for proper management of water resources in a given area. This can lead to the oft-occurring phenomenon of deterioration in the quality of groundwater at shallow depths that may occur either alone or as a combination of physical or chemical, and at some locations, radioactive contamination. Conventional approaches for studying the groundwater quality got impetus with the advent of tracer techniques involving the applications of stable and/or radioisotopes in soil-water systems. In this regard, spatial and temporal variations of environmental stable isotopes of water (measured as δ^2H and $\delta^{18}O$) from different sources have been used to solve certain hydrological problems which are, at times, not possible to study by conventional methods. Accordingly, the present paper deals with the analysis of variations of stable isotopes and environmental tritium contents of water samples vis-à-vis corresponding chemical parameters to study the interconnection between water sources at three locations namely, Jodhpur city, Jamarkotra mine of Udaipur district, Indira Gandhi Nahar Command (IGNC) area in Rajasthan. Studies revealed that the source of seepage in some buildings of Jodhpur city could either be directly from the Kailana lake or indirectly through supply lines to the city, there has no been evidence of any contribution of reservoir waters to the groundwater in Jamarkotra mine area, and water logging and salinity problems could be due to over irrigation and presence of gypsiferous layers at varying depths at Lunkaransar farm, IGNC, Rajasthan. Results are discussed in the light of secondary information and possible recommendations have been made.

Key Words: Stable Isotopes, interconnection, salinization, seepage, gypsiferous layer.

Introduction

The stable isotopes contents (measured as δ^2H and $\delta^{18}O$) and environmental ^3H of water from different sources are known to vary with location and/or time. Stable isotope variations are attributed to the differential behaviour of water molecules subjected to natural processes like precipitation, salinization, evaporation, intermixing, absorption and transpiration of crops, etc. Hence, changes in the stable isotopes coupled with the environmental tritium contents of different water samples can potentially be used to understand hydrological processes in a given area (Gonfiantini et al. 1974; IAEA, 1981; Allison 1982; Allison et al. 1983; Bhattacharya et al. 1985; Navada et al. 1986; Chandrasekharan et. al. 1988; Datta et al. 1991).

Based on δ^2H, $\delta^{18}O$ and chloride contents of shallow groundwater samples drawn through auger holes in Algeria, it was concluded that evaporation was the main mechanism of salinization process in that area (Gonfiantini et al. 1974). The role of climate and vegetation through transpiration affecting stable isotope contents was studied by Allison et al. (1983). The contribution of irrigation water and rainfall to the groundwater has been analysed by stable isotopic analytical approach in Pakistan (Haq et al. 1983). Characteristics of stable isotopes vis-à-vis land use pattern has been discussed and reported by Chandrasekharan et. al. (1992) and Sundara Sarma et al. (1993).

In this paper, isotope investigations for interconnection between water bodies carried out at three locations in Rajasthan and one in Gujarat are presented and results are discussed to evolve specific recommendations.

Areas of investigation

Jodhpur city: This city was primarily designed to arrest rainwater in impounding structures to provide sustained water supply to the populace. The city consists of a number of wells and step wells. The lift canal from Indira Gandhi Nahar is used to fill the Kailana Lake the main source of water which is for supply domestic purposes. Problems associated with seepage water accumulation in the basement of a number of buildings in some parts of the city and rise in static water level were reported in 1998. A schematic representation of the investigated area is shown (figure 1).

Jhamarkotra mine, Udaipur: This mine is situated at about 25 km. Southeast of Udaipur city, Jhamarkotra mine is a mechanised open cast for phosphate rock extraction and exists for over 32 years. This area is surrounded by few man-made reservoirs within nine km. radius. The existence of reservoirs and minimal decrease in drawdown became a cause of anxiety that could be due to the ingress of reservoir water into mine area.

Lunkaransar farm, IGNC: This study area is located at about one and half kilometre southeast of the lift canal and about 72 km north of Bikaner, the nearest city (figure 2). The farm consists of 54 plots, each of 1.5 ha area (except the corner ones). These plots were provided with irrigation facilities through three main water channels drawn from the lift canal. It is reported that prior to canal irrigation, the entire farm area was comprised of sand dunes and there were no perched water or wells of any type. With the advent of canal irrigation, dunes were leveled and agricultural practices were initiated in selected plots. Occurrence of perched water in the farm were noted after continuous irrigation in cultivated plots and now it has

been established that the gypsiferous layers at shallow depths are responsible for the occurrence of perched water in the farm area (Das *et al.* 1992; Chandrasekharan *et al.* 1992).

Present investigations

Water samples from different sources from the locations mentioned above were collected for stable isotopes and chemical analyses. Some water samples from Jodhpur and Jhamarkotra mine areas were collected for the study of environmental tritium contents. While stable isotope contents were measured in Micro-mass spectrometer, tritium contents were measured using Liquid Scintillation Counter at BARC, Mumbai. Hydro-chemical analyses of samples from IGNC were done using conventional methods. Geo-electrical investigations were conducted at IGNC area to understand the subsurface configuration and help in interpreting the isotopic data.

Results and discussion

Jodhpur area: Measured δ^2H and $\delta^{18}O$ and environmental tritium contents of water samples from different sources are presented in Table 1. Plot of δ^2H and $\delta^{18}O$ shows that the water samples from lake and filter houses (which have been supplied by lake) are depleted compared to other surface water bodies (fig. 3). Basement samples fall between lake water and groundwater, which suggests contribution of lake water to the basement seepage water (IAEA, 1983). Pond water samples are highly enriched in δ^2H and $\delta^{18}O$ showing evaporation effect. Tritium content of lake water and filter houses water vary from 9 to 12 TU whereas basement samples show tritium content of 6.5 to 10 TU (figure 3). Tritium content of hand pump samples is in the range of 2.5 to 7 TU (Nair *et al.* 1993). Tritium results also suggest that basement samples are mixture of lake water and groundwater. Spring sample from Vyas park shows depleted value of δ^2H and $\delta^{18}O$ (more depleted than basement samples) and higher tritium content (higher than basement samples)

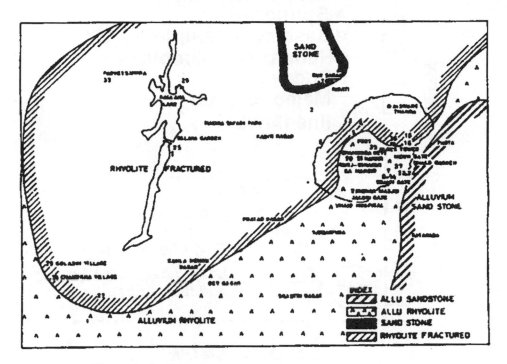

Figure 1: Geological map of Jadhpur city and surrounding areas

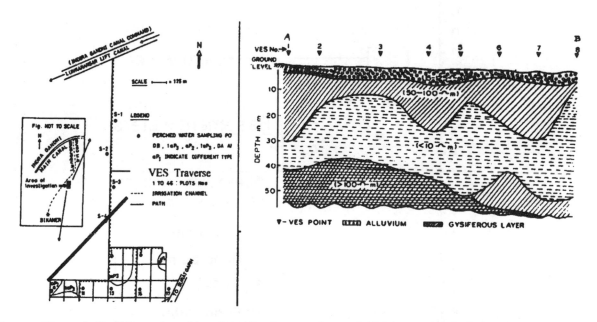

Figure 2: (Left) Investigated location at Lunkaransar farm and (right) geo-electrical along a traverse

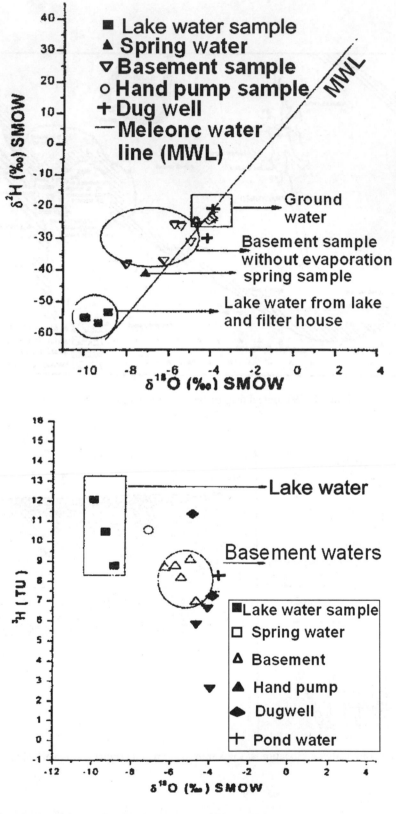

Figure 3: Plots showing variations of stable isotopes and tritium contents of water samples from Jodhpur areas

indicating a good component of lake water. Direct seepage from Kailana lake (RL 280m) to Vyas Park spring (RL310m) is not possible due to higher elevation of Vyas Park. Most probably the used water percolating to the subsurface is contributing to the spring.

Jhamarkotra mine, Rajasthan: In general, the electrical conductivity of samples from the mine area is around 0.68 dS m^{-1} whereas for the groundwater samples it ranges from 0.92 dS m^{-1} to 2.35 dS m^{-1}. EC values for surface water reservoirs of Jhamri, Jaisamand and Uday Sagar have been 0.5 – 0.55 dS m^{-1} and 1.0–1.2dSm^{-1} respectively. Thus there is marked variation in EC values of water samples from different sources.

It could be observed that highly enriched isotopic content exhibited by surface waters compared to that of ground water samples (Table 2). The isotopic composition of surface water shows considerable fluctuations during pre- and post- monsoon samples whereas isotopic composition of groundwater in the dolomitic limestone falls in a fairly constant range of –3 ‰ -6 ‰ irrespective of season. These suggest that there is no ingress or recharge from surface water bodies to the ground waters in the mine area. The scatter of groundwater samples on this plot is mainly exhibited by samples from shallow dugwells and handpumped wells from phyllitic and gneissic (BGC) aquifers. This scatter is attributed to evaporation during summer.

Plot of $\delta^{18}O$ – tritium suggests that the possibility of influence of surface waters from reservoirs on groundwater in the mine area is remote (Fig. 4). Most of the wells in dolomitic limestone in the mine area show tritium content below 5TU. A few dug wells and shallow hand pumped wells as well as surface water samples show tritium levels from about 6TU to 10 TU. Mixing of modern and comparatively older water is also apparent. This could be due to continuous pumping of water from the aquifer. In most of the tube wells in the mine, $\delta^{18}O$ enrichment is observed with time due to continuous pumping. The $\delta^{18}O$ enrichment is interpreted as recharge to the mine area either by percolation of mine discharge through dolomitic limestone or by back pressure through the pumping wells.

Lunkaransar farm, IGNC: δ^2H -- $\delta^{18}O$ plot yielded the following relationship

$$\delta^2H = 6.0 \delta^{18}O - 1.72 : (n=18, r=0.87).$$

The slope 6.0 indicates the effect of evaporation on δ-values. Further analyses of δ-lines under different land use conditions yield different slopes (m) and correlation coefficients as shown (Table 3). Plot of δ^2H and $\delta^{18}O$ values for IGNC and other areas are shown in fig. 5. The above values show that the slope of δ-line is nearly parallel to that of meteoric line in the case of samples from cropped plots and the slope decreased with decreasing cropland use which is followed by increased evaporation from the perched water. The wide variation of δ-values within a particular land use may be due to the mixing of canal water with the perched water in the form of irrigation.

The mechanism of salnization in the farm area was evaluated (Chandrasekharan *et al.* 1992). Accordingly, salinization in the farm could be caused by one or a combination of the following processes:

• Leaching of salts in cultivated plots (applicable to cropped and partial fallow plots),
• High evaporative conditions and subsequent upward movement of salts from the highly saline zone below (Sundara Sarma *et al.* 1993), the gypsiferous layers (mainly applicable to fallow and partial fallow plots) and
• Subsurface flow of irrigation excess water in lateral and vertical directions (applicable to all plots).

With these possibilities, within fallow plots, there has been a large difference of $\delta^{18}O$ values and electrical conductivities (in the range of 2.32 – 47.5 dS m^{-1}) indicating that evaporation may not be the only factor for variability of δ-values and salinity. Reasons for scattering of points

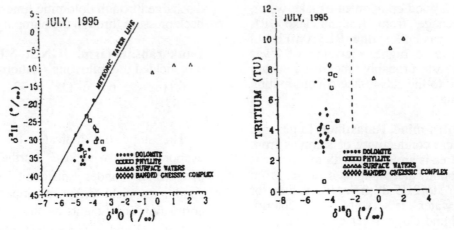

Figure 4: Plots showing variation of stable isotopes and tritium contents of water samples from Jhamarkotra mine area, Udaipur, Rajasthan

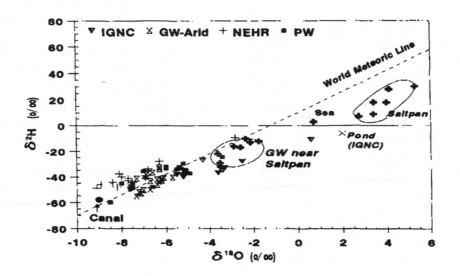

Figure 5: Variation of stable isotope contents of water from different sources

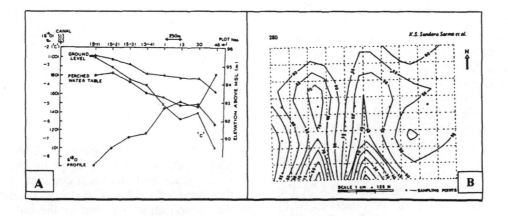

Figure 6: (A) Profile of δ^2O and EC of perched water, water table from canal- plot46 (B) Contribution of canal water (as %) to perched water at Lunkaransar farm, Rajasthan

Table- 1 : δ^2H, $\delta^{18}O$ and Environmental Tritium Contents of Samples from Different Sources of Water in Jodhpur City

SNo.	Location	δ^2H	$\delta^{18}O$	Env. ^3H
	LAKE SAMPLES			
1	Kailana Lake	-56.6	-9.31	10.5
2	Fetehsagar reservoir	-53.3	-8.86	8.8
3	Kailana filterhouse	-55.1	-9.89	12.1
	SPRING SAMPLE			
4	Vyas Park	-41.1	-7.14	10.1
	BASEMENT SAMPLES			
5.	Unique Traders	-30.8	-4.95	9.1
6.	Surana Market	-36.8	-6.25	8.7
7.	Kamla Market	-38.1	-8.06	-
8.	Mayur Arwwl	-21.9	-4.48	8.2
9.	Moti Chowk	-25.5	-4.17	9.7
10.	Kunj B. Mandir	-25.6	-5.72	8.8
	HANDPUMP SAMPLES			
11.	Opp. Gorinda Baori	-24.8	-5.64	6.3
12.	Moti Chowk	-23.7	-3.92	2.7
13.	Katla Bazaar	-24.3	-3.81	6.7
14.	Nai Sadak	-24.8	-4.15	5.9
	DUG WELL SAMPLES			
15.	Dab Garon Kua	--	-4.88	11.4
16.	Moti Chowk	-25.5	-3.79	7.3
17.	Chopasni	-21.9	-3.81	6.3
18.	Devnagar	-29.8	-4.15	--
	POND SAMPLES			
19.	Ranisar pond	-22.1	-3.51	8.3
20.	Chand Baori	-25.7	-4.02	7.4
21.	Krva Jhalra	-29.8	-3.61	7.4
22.	Gorind Baori	-23.5	-4.34	6.9

Table- 2: δ^2H, $\delta^{18}O$ and Environmental Tritium Contents of Samples from Different Sources of Jhamarkotra Mine Area, Udaipur

SNo.	Location	δ^2H	$\delta^{18}O$	Env. ^3H
	DOLOMITIC LIMESTONE			
1	JK1	-37.0	-4.60	4.1
2	JK2	32.0	-4.50	3.2
3	JK3	-35.0	-4.60	5.0
4	JK4	-33.0	-4.50	4.3
5	JK5	-37.0	-4.70	4.1
6	JK6	-37.0	-4.40	2.5
7	JK7	-35.0	-4.10	3.3
8	JK8	-35.0	-4.40	2.8
9	JK9	-28.9	-4.90	1.6
10	JK10	-28.1	-4.00	3.5
11	JK14	-31.0	-4.20	3.7
12	JK17	33.0	-4.60	3.6
13	JK18	-22.1	-2.80	-
14	JK20	-36.0	-5.00	3.1

15	JK23		-34.0	-4.36	0.6
16	JK24		-33.0	-4.70	4.0
17	JK29		-36.0	-4.50	3.1
18	JK31		-33.0	-3.30	4.5
19	JK32		-31.7	-4.50	5.2
20	JK33		-33.0	-3.20	4.5
21	JK35		-29.0	-4.90	7~.1
22	JK36		-30.7	-4.00	4.9
23	JK37		-28.6	-5.10	7.5
24	JK38		-33.0	-4.10	5.3
	PHYLLITE				
25	JK12		-25.0	-4.00	4.1
26	JK15		-27.9	-3.60	6.7
27	JK16		-23.9	-4.20	--
28	JK28		-31.0	-3.30	7.5
29	JK39		-31.0	-3.70	7.6
	GRANITE AND GNEISES				
30	JK27		-27.0	-3.75	8.2
	RESERVOIRS				
31	JK13		-12.0	-0.20	7.3
32	JK21		-10.0	+2.20	9.9
33	JK30		-10.0	+1.16	9.3

Table 3: $\delta^2 H$--$\delta^{18} O$ Relations for Different Land Use Conditions at Lunkaransar Farm, Western Rajasthan

Land use	Slope (m)	Intercept (d)	No of Samples	Correlation Coefficient (r)
Cropped	8.61	16.51	7	0.96
Partial Fallow	5.18	- 7.47	7	0.66
Fallow	4.54	- 8.73	4	0.99

corresponding to partial fallow and cropped plots may be attributed to the application of varying amount of agro-chemicals and irrigation at different times in the cultivated plots. Thus it could be possible that the salinity in perched water is affected by land use conditions besides the seepage of saline water through the gypsiferous layer at shallow depth and at the same time the variation may be governed by the amount of canal water (as irrigation excess) mixing. On this basis, the contribution of irrigation water to perched water has been calculated by a simple isotope mass balance relationship and the same is presented in the form of contours (fig. 6).

Concluding remarks

On the basis of geohydrological investigations (comprising isotope, hydrochemical, geophysical and hydrogeological techniques) the following points and suggestions could emerge:

- **Jodhpur Area:** The Kailana lake water is contributing to the seepage water in the basement of some of the buildings in the city. Absence of E-coli in the entire basement samples rules out the possibility of seepage from sewer lines. The lake water contribution to the seepage could either be due to direct seepage from Kailana lake or seepage from pipelines and used water percolating to the subsurface. Increased consumption of Lake Water and discontinuation of groundwater withdrawal have further aggravated the seepage problem.

- **Jhamarkotra Mine, Rajasthan:** There is no evidence of any contribution from reservoir to the groundwater in the mine area. The groundwater in the mine is mainly replenished by precipitation. Some of the groundwater pumped out of the aquifer is percolating back to aquifer

100

either through dolomitic limestone or through one or more pumping wells due to back pressure. Increased pumping is drawing older water out of the aquifer. Installtion of tube wells on the periphery of mine would widen the cone of depression and result in lowering of the water table in the mine area. It would also help to stop the inflow of modern recharge to the mine.

- **Lunkaransar farm, IGNC:** δ-values of perched water in the farm were affected by land use and salinization which is dominant in fallow plots. The amount of canal water (as irrigation excess) in percentage computed using the simple isotope mass balance relation indicates the deterioration in the quality of perched water away from the irrigation channels.

The above case studies illustrate potentialities of isotope techniques in conjunction with conventional methods to solve specific hydrological problems not answerable by conventional methods alone. It is suggested that similar approach be followed in such areas for the overall development of water resources in the country.

Acknowledgement

Authors are grateful to the Director, IARI, New Delhi and the Project Director, Water Technology Centre, IARI for kindly providing facilities. The help rendered by colleagues at IARI and BARC in completing case studies is duly acknowledged.

References

Allison G.B., 1982 *Journal of Hydrology,* **55** (1/4): 163-169.

Allison G.B, Barnes C.J., Huges M.W. and Leaney Fwj, 1983 *Proc. International Symp. on Isotope Hydrology,* IAEA,Vienna, 12-16, Sept.: 105-124.

Bhattacharya S.K., Gupta S.K. and Krishnamoorthy R.V., 1985 Proc. *Indian Acad. of Sciences (Earth and Planetary Sciences)* **94**, 3, Nov.: 283-295.

Chandrasekharan H., Navada S.V., Jain S.K., Rao S.M. and Singh Y.P., 1988 *Natural Recharge of Ground water; NATO-ASI Series on Mathematical and Physical Sciences,* Amsterdam, The Netherlands, **222**: 202-220.

Chandrasekharan H., Sundara Sarma K.S., Das D.K., Dutta D, Mookerjee and Navada S.V., 1992 *Journal of Arid Environments,* 23: 365-378.

Craig H., 1961 *Science,* **133**, 1833-1834.

Das D.K., Sundara Sarma K.S., Chandrasekharan H., Datta D., Saha B. and Mookerjee P., 1992 *Technical Report* submitted to: Department of Science and Technology, Govt. of India, New Delhi; p. 177.

Das P.K., Kakkar YP,Moser H and Stichler W 1988 *Journal of Hydrology,* 98: 133-146.

Datta P.S., Tyagi S.K. and Chandrasekharan H., 1991 *Journal of Hydrology,* **128**: 223-236.

Gonfiantini R., Dincer T. and Derekoy A.M., 1974 *Proc. International Symp. on Isotope Techniques in Groundwater Hydrology,* IAEA, Vienna, 11-15, March, **I**: 293-388.

GWD 1998 *Technical Report,* Groundwater Department, Jodhpur, Rajasthan.

Haq M.I., Sajjad M.I. and Malik K.A., 1983 *Proc. International Symposium on Isotope and Radiation Techniques in Soil Physics and Irrigation Studies;* IAEA,Vienna: 375-388.

IAEA 1981 *Technical Report,* Series No.**210**: 334.

Nair A.R., Sinha U.K., Joseph TB and Navada SV 1993 *Proc. 2nd National Symposium on Environment,* Jodhpur: 188-190.

Navada S.V., Jain SK, Shivanna K. and Rao S.M. 1986 *Indian Journal of Earth Sciences,* **13**, 2: 223-234.

Sundara Sarma K.S,, ChandrasekharanH, Navada SV, Dutta D, Mookerjee P and Das DK 1993 *IAHS Publication,* **215**: 275 281.

MODELLING

Intl. Conf. on Sustainable Development and Management of Groundwater Resources
in Semi-Arid Region with Special Reference to Hard Rock, (IGC 2002),
M. Thangarajan, S.N. Rai & V.S. Singh (Eds.)

Hybrid finite analytic solution for stream aquifer interaction

A. Upadhyaya[1], H.S. Chauhan[2] and S.R. Singh[1]

[1]ICAR Research Complex for Eastern Region, WALMI Complex, Patna
[2]Dept. of Irrigation and Drainage Engg., G.B.P.U.A.T., Pantnagar

Abstract

Water table varies in a semi-infinite aquifer due to its interaction with the stream/canal having suddenly rising or declining water levels. Transient water table profiles in a recharging and discharging aquifer is predicted for a selected numerical example by employing hybrid finite analytic method. In this method nonlinear Boussinesq equation is locally linearized and solved analytically after approximating the unsteady term by a simple finite difference formula to approximately preserve the overall nonlinear effect by the assembly of local analytic solutions. Water table profiles at t =1 day and at t = 5 days as obtained by this solution are compared with those obtained by employing finite element method. There is no significant difference in the values of water table profile computed by both the methods. The proposed method is quite simple as compared to finite element method so this may be used for studying water table variation due to stream aquifer interaction.

Key words: Semi-infinite flow, transient, hydrid finite analytic solution, linearized Boussinesq equation, recharging, discharging aquifers

Introduction

Interaction of a stream or canal with the aquifer is a natural phenomenon and ground water levels in the aquifer fluctuate in response to the abrupt rise or fall of water level in the associated stream. If water level in the interacting stream is more than the water level in the aquifer, the stream will recharge the aquifer. Similarly if water level in the stream is less than the aquifer water level, the aquifer will discharge into the stream. Thus the process of recharging or discharging of aquifer, which is interacting with the stream, is governed by the change in gradient between stream and aquifer. The rate of change of water level in the aquifer will depend upon the transmissivity and the porosity of the aquifer, and the change in stream water level.

Many investigators have studied the water table variation in a semi infinite horizontal aquifer due to sudden rise or fall of water level in the adjoining stream by obtaining and employing analytical or numerical solution of one-dimensional Boussinesq equation. Such studies include those by Marino (1973), Sidiropoulos et al. (1984), Tolikas et al. (1984), Lockington (1997), Workman et al. (1997), Serrano and Workman (1998), and Upadhyaya and Chauhan (1998).

Further, if the impermeable barrier on which aquifer is resting, is inclined with a mild slope, it influences the water table variation in the aquifer. Only a few studies seem to be related to stream and sloping aquifer interaction. Polubarinova-Kochina (1962) obtained an analytical solution of the linearized Boussinesq equation to describe seepage from one canal to another on sloping bedrock. Yussuff et al. (1994) obtained a finite difference numerical solution of the nonlinear Boussinesq equation characterizing the phreatic surface in a semi-infinite sloping aquifer. They also obtained an analytical solution by modifying Polubarinova-Kochina's solution (1962) of a generalized boundary condition to describe seepage from a canal in a semi-infinite flow region. They observed that phreatic surface values predicted by the numerical solution were overall higher for all distances and times than those predicted by the analytical solution of the linearized Boussinesq equation. Upadhyaya (1999) obtained various analytical, numerical, and hybrid finite analytic solutions of one - dimensional Boussinesq equation to study water table variation in a number of physically identified flow situations in a horizontal and sloping aquifers. Upadhyaya and Chauhan (2001a) obtained an analytical solution of the linearized Boussinesq equation and a finite-element numerical solution of the nonlinear Boussinesq equation to describe water table variation in a sloping/ horizontal aquifer receiving constant replenishment and interacting with the adjoining stream having an abrupt rise or drop of water level. Such problem can also be solved by employing hybrid finite analytic approach presented by Chen (1988). In this approach the non-linear partial differential equation is locally linearized and solved analytically in space and discretized in time by a simple difference formula. The resultant system of algebraic equations approximate the overall nonlinear effect because some non-linear terms are treated as constants only in the local regions. A four point numerical formula provides stable and sufficiently accurate results with simple calculations. Pi and Hjelmfelt (1994) solved an extended Dupuit-Forchheimer equation to describe water table variation and lateral subsurface storm flow in a sloping finite aquifer using hybrid finite analytic approach and showed that steady state profiles of water table and lateral subsurface storm flow obtained from their study compared well with the results of previous investigators. Upadhyaya and Chauhan (2001b) compared the falling water tables between two drains as predicted by various analytical and numerical solutions of Boussinesq equation with those obtained from the experimental solution on vertical Hele-shaw model and concluded that hybrid finite analytic solution predicts the values of falling water table at mid point closest to the values obtained from experimental solution. Since hybrid finite analytic approach seems better and no studies are available in literature related to solution of one dimensional Boussinesq equation representing stream and sloping aquifer interaction based on this approach,

so hybrid finite analytic solution is obtained for such situation in the present study and results are compared with those predicted by existing finite element numerical solution. A brief description of boundary value problem and hybrid finite analytic approach is presented below.

Problem formulation

The phenomenon of water table variation in a sloping semi-infinite aquifer receiving constant replenishment and interacting with the stream having a sudden rise or fall of water can be represented by one-dimensional non-linear Boussinesq equation with appropriate initial and boundary conditions and written as:

$$h\frac{\partial^2 h}{\partial x^2}+\left(\frac{\partial h}{\partial x}\right)^2-\alpha\left(\frac{\partial h}{\partial x}\right)+\frac{R}{K}=\frac{f}{K}\left(\frac{\partial h}{\partial t}\right) \quad (1)$$

$$h = h_1; x = 0; t > 0 \quad (2)$$
$$h = h_0; x > 0; t = 0 \quad (3)$$
$$h = h_0; x \rightarrow 0; t > 0 \quad (4)$$

where h = height of the phreatic surface above the sloping impermeable barrier [L]; α = slope of the impermeable barrier; x = space coordinate along horizontal axis [L]; t = time [T]; K = hydraulic conductivity of aquifer [LT^{-1}]; f = drainable porosity (dimensionless); and R = surface applied replenishment [LT^{-1}]; h_1 and h_0 denote water levels in the stream at x = 0 and in the aquifer at x = ∞. The definition sketches of the water table profile in recharging and discharging aquifers are given below in Figs 1 and 2, respectively.

Hybrid finite analytic solution

Using the approach of Chen (1988) the hybrid finite analytic solution of nonlinear equation (1) along with initial and boundary conditions (2-4) was obtained. The mathematical steps are given below.

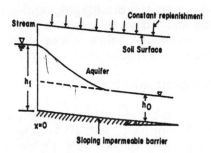

Figure 1: Recharging aquifer with constant replenishment from land surface

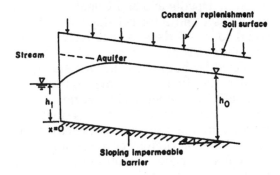

Figure 2: Discharging aquifer with constant replenishment from land surface

In order to absorb the terms of constant recharge in eq (1) a transformation is devised as:

$$h = v + \frac{Rt}{f} \qquad (5)$$

This transformation converts eq (1) into the following form:

$$\frac{\partial^2 v}{\partial x^2} + \frac{1}{D}\left(\frac{\partial v}{\partial x}\right)^2 - 2s\left(\frac{\partial v}{\partial x}\right) = \frac{1}{a}\frac{\partial v}{\partial t} \qquad (6)$$

where D is average depth of flow; s = $\alpha/2D$; and a = KD/f.

Assuming the terms $1/D\ (\partial v/\partial x)^2$ and $1/a\ ((\partial v/\partial t)$ as constants in a small sub region and denoted by C and E, respectively, the integration yields following equation:

$$\frac{dv}{dx} - 2sv - (E - C)x + F \qquad (7)$$

The solution of this first order ordinary differential equation is:

$$v(x) = G\,e^{2sx} - (E - C)\frac{x}{2s} + I \qquad (8)$$

Discretization of eq (8) in space and time yields following equations.

$$v_i^{n+1} = G + I \qquad (9)$$

$$v_{i-1}^{n+1} = G\,e^{-2s\Delta x} + (E - C)\frac{\Delta x}{2s} + I \qquad (10)$$

$$v_{i+1}^{n+1} = G\,e^{2s\Delta x} - (E - C)\frac{\Delta x}{2s} + I \qquad (11)$$

because v_i^{n+1} represents the point where $\Delta x = 0$.

Simplification of these equations yield a tridiagonal matrix as:

$$A_i v_{i-1}^{n+1} + B_i v_i^{n+1} + C_i v_{i+1}^{n+1} = D_i v_i^n + E_i \qquad (12)$$

where

$$A_i = -\frac{e^{s\Delta x}}{e^{s\Delta x} + e^{-s\Delta x}} \qquad (13)$$

$$B_i = 1 + \frac{e^{s\Delta x} + xe^{-s\Delta x}}{2as\Delta t} \qquad (14)$$

$$C_i = -\frac{e^{-s\Delta x}}{e^{s\Delta x} + e^{-s\Delta x}} \qquad (15)$$

$$D_i = \frac{\Delta x}{2as\Delta t}\tanh\,(s\Delta x) \qquad (16)$$

$$E_i = \left(\frac{\partial v}{\partial x}\right)^2 \frac{\Delta x}{2sD}\tan h(s\Delta x) \qquad (17)$$

By solving the tridiagonal matrix in eq (12) the value of v at different nodes can be obtained. Again by applying inverse of transformation (5) the value of h at a given space and time may be computed.

It should be possible to obtain water table variation in a horizontal aquifer by putting s = 0 in the above solution but then the expression becomes indeterminate. However, water table elevations in a horizontal aquifer may be obtained by substituting a very small value of α such as 0.00001 in the above solution without affecting its general stability.

Results and discussion

Spatial and temporal variation of water table in a sloping semi-infinite aquifer as a result of abrupt change of water level in the adjoining stream was evaluated by considering a numerical example. The effect of constant replenishment from the land surface on water table in both the recharging and discharging aquifers was also evaluated. The same numerical example, which was considered for finite element solution, has been used to get the results. The example is reproduced again for ready reference.

Assume the flow of water in a shallow sand aquifer with hydraulic conductivity K = 20 m/day and specific yield f =0.27. The aquifer was considered to be underlain by an impermeable barrier having 0, 5 and 10% slopes and initially having a uniform water level h_0 = 2 m. The water level in the adjoining trench was instantaneously raised to the elevation h_1 = 3 m to provide an interacting recharging aquifer. Similarly when the water level in the aquifer was at an elevation of h_0 = 3 m, and in the canal/ stream it was at an elevation of h_1 = 2 m, it provides an interacting discharging aquifer. A constant replenishment of 5 mm/day was assumed. The values of time increment Δt and space increment Δx were considered as 0.0025 days and 2 m, respectively.

Water table variation in a horizontal/ sloping recharging and discharging aquifers on account of abrupt rise or fall of water level in the adjoining stream was computed by employing hybrid finite analytic solution. The water table heights in recharging and discharging aquifers at t = 1 day and 5 days are given in Tables 1-8 and presented below separately for recharging and discharging aquifers.

Recharging aquifer

Water table elevations in a recharging aquifer having 0, 5, and 10 % slopes and receiving zero or constant recharge are computed by hybrid finite analytic solution and are compared with the results obtained from finite element solution given by Upadhyaya and Chauhan (2001a) in Tables 1 and 2 for t=1 day and Tables 3 and 4 for t = 5 days, respectively. It may be observed that water table heights predicted by both the solutions are almost similar and finally become constant and parallel to the impermeable barrier. It may also be seen from Tables 1-4 that with increase in slope of the impermeable base the water table height corresponding to a particular space coordinate (except boundaries) as obtained from both the solutions increases. Comparison of water table heights at t = 1 day and t = 5 days (without replenishment in Tables 1 and 3 and with constant replenishment in Tables 2 and 4) shows that with the increase in time water table elevation at a particular space coordinate increases. At t = 5 days the water table profile becomes parallel to impermeable base at larger distances compared to t = 1 day. Effect of a constant replenishment of 5 mm/day in case of recharging aquifer on water table at t = 1 day and 5 days can be observed from the fact that the values of water table elevation obtained from both the solutions are larger (in Tables 2 and 4) than those values of water table elevations in the absence of replenishment given (in Tables 1 and 3).

Discharging aquifer

Water table elevations in a discharging aquifer having 0, 5, and 10 % slope and receiving zero and constant replenishment are given in Table 5 and 6 respectively for t = 1 day, and in Table 7

Table 1. Comparison of water table heights for t = 1 day predicted by Hybrid Finite Analytic Solution (HFAS) and Finite Element Solution (FES) for horizontal and sloping recharging aquifer without replenishment

X (m)	0 % slope		5 % slope		10 % slope	
	HFAS	FES	HFAS	FES	HFAS	FES
0.0	3.000	3.000	3.000	3.000	3.000	3.000
10.0	2.637	2.638	2.694	2.694	2.746	2.746
20.0	2.317	2.318	2.387	2.387	2.458	2.458
30.0	2.116	2.116	2.161	2.161	2.215	2.215
40.0	2.031	2.030	2.048	2.048	2.074	2.073
50.0	2.006	2.005	2.011	2.010	2.018	2.018
60.0	2.001	2.001	2.002	2.002	2.003	2.003
70.0	2.000	2.000	2.000	2.000	2.000	2.000
80.0	2.000	2.000	2.000	2.000	2.000	2.000

Table 2: Comparison of water table heights for t = 1 day predicted by Hybrid Finite Analytic Solution (HFAS) and Finite Element Solution (FES) for horizontal and sloping recharging aquifer with constant replenishment equal to 5 mm/day

X (m)	0 % slope		5 % slope		10 % slope	
	HFAS	FES	HFAS	FES	HFAS	FES
0.0	3.000	3.000	3.000	3.000	3.000	3.000
10.0	2.648	2.648	2.705	2.704	2.756	2.755
20.0	2.334	2.334	2.404	2.402	2.475	2.473
30.0	2.135	2.134	2.181	2.179	2.236	2.233
40.0	2.050	2.048	2.068	2.066	2.094	2.092
50.0	2.025	2.024	2.030	2.029	2.038	2.036
60.0	2.020	2.019	2.020	2.020	2.022	2.022
70.0	2.019	2.019	2.019	2.019	2.019	2.019
80.0	2.019	2.019	2.019	2.019	2.019	2.019

Table 3: Comparison of water table heights for t = 5 days predicted by Hybrid Finite Analytic Solution (HFAS) and Finite Element Solution (FES) for horizontal and sloping recharging aquifer without replenishment

X (m)	0 % slope		5 % slope		10 % slope	
	HFAS	FES	HFAS	FES	HFAS	FES
0.0	3.000	3.000	3.000	3.000	3.000	3.000
10.0	2.838	2.838	2.901	2.901	2.944	2.944
20.0	2.675	2.675	2.786	2.786	2.872	2.872
30.0	2.518	2.518	2.662	2.662	2.783	2.783
40.0	2.376	2.377	2.534	2.534	2.681	2.682
50.0	2.256	2.258	2.410	2.410	2.571	2.571
60.0	2.164	2.165	2.297	2.297	2.458	2.458
70.0	2.099	2.098	2.203	2.203	2.349	2.349
80.0	2.056	2.055	2.129	2.129	2.251	2.251
90.0	2.030	2.029	2.077	2.077	2.170	2.170
100.0	2.015	2.014	2.043	2.043	2.108	2.107
110.0	2.007	2.006	2.022	2.022	2.064	2.064
120.0	2.003	2.003	2.011	2.011	2.035	2.035
130.0	2.001	2.001	2.005	2.005	2.018	2.018
140.0	2.000	2.000	2.002	2.002	2.009	2.009
150.0	2.000	2.000	2.001	2.001	2.004	2.004
160.0	2.000	2.000	2.000	2.000	2.002	2.002
170.0	2.000	2.000	2.000	2.000	2.001	2.001
180.0	2.000	2.000	2.000	2.000	2.000	2.000

Table 4: Comparison of water table heights for t = 5 days predicted by Hybrid Finite Analytic Solution (HFAS) and Finite Element Solution (FES) for horizontal and sloping recharging aquifer with 5 mm/day replenishment

X (m)	0 % slope		5 % slope		10 % slope	
	HFAS	FES	HFAS	FES	HFAS	FES
0.0	3.000	3.000	3.000	3.000	3.000	3.000
10.0	2.864	2.864	2.922	2.921	2.962	2.961
20.0	2.722	2.722	2.826	2.825	2.905	2.904
30.0	2.582	2.582	2.717	2.716	2.831	2.828
40.0	2.453	2.453	2.602	2.600	2.742	2.738
50.0	2.341	2.342	2.488	2.486	2.642	2.638
60.0	2.253	2.254	2.383	2.381	2.538	2.533
70.0	2.190	2.191	2.294	2.291	2.436	2.431
80.0	2.148	2.148	2.223	2.221	2.343	2.338
90.0	2.122	2.122	2.172	2.170	2.265	2.261
100.0	2.107	2.107	2.138	2.136	2.204	2.200
110.0	2.100	2.099	2.117	2.116	2.160	2.157
120.0	2.096	2.096	2.105	2.104	2.131	2.129
130.0	2.094	2.094	2.098	2.098	2.113	2.112
140.0	2.093	2.093	2.095	2.095	2.103	2.102
150.0	2.093	2.093	2.094	2.094	2.097	2.097
160.0	2.093	2.093	2.093	2.093	2.095	2.094
170.0	2.093	2.093	2.093	2.093	2.093	2.093
180.0	2.093	2.093	2.093	2.093	2.093	2.093

Table 5: Comparison of water table heights for t = 1 day as predicted by Hybrid Finite Analytic Solution (HFAS) and Finite Element Solution (FES) for horizontal /sloping discharging aquifer without replenishment

X (m)	0 % slope		5 % slope		10 % slope	
	HFAS	FES	HFAS	FES	HFAS	FES
0.0	2.000	2.000	2.000	2.000	2.000	2.000
10.0	2.432	2.432	2.372	2.371	2.312	2.312
20.0	2.717	2.717	2.660	2.660	2.597	2.598
30.0	2.881	2.878	2.841	2.842	2.798	2.799
40.0	2.958	2.955	2.937	2.938	2.914	2.915
50.0	2.988	2.986	2.979	2.980	2.969	2.970
60.0	2.997	2.997	2.994	2.994	2.991	2.991
70.0	2.999	2.999	2.999	2.999	2.998	2.998
80.0	3.000	3.000	3.000	3.000	3.000	3.000
90.0	3.000	3.000	3.000	3.000	3.000	3.000

Table 6: Comparison of water table heights for t = 1 day as predicted by Hybrid Finite Analytic Solution (HFAS) and Finite Element Solution (FES) for horizontal /sloping discharging aquifer with constant replenishment equal to 5 mm/day

X (m)	0 % slope		5 % slope		10 % slope	
	HFAS	FES	HFAS	FES	HFAS	FES
0.0	2.000	2.000	2.000	2.000	2.000	2.000
10.0	2.441	2.444	2.380	2.383	2.319	2.322
20.0	2.730	2.733	2.671	2.675	2.607	2.613
30.0	2.896	2.895	2.855	2.859	2.811	2.816
40.0	2.975	2.973	2.953	2.956	2.929	2.933
50.0	3.006	3.005	2.996	2.998	2.986	2.988
60.0	3.015	3.015	3.012	3.013	3.000	3.009
70.0	3.018	3.018	3.017	3.017	3.016	3.016
80.0	3.018	3.018	3.018	3.018	3.018	3.018
90.0	3.018	3.018	3.018	3.018	3.018	3.018

Table 7: Comparison of water table heights for t = 5 days as predicted by Hybrid Finite Analytic Solution (HFAS) and Finite Element Solution (FES) for horizontal /sloping discharging aquifer without replenishment

X (m)	0 % slope		5 % slope		10 % slope	
	HFAS	FES	HFAS	FES	HFAS	FES
0.0	2.000	2.000	2.000	2.000	2.000	2.000
10.0	2.212	2.212	2.128	2.128	2.066	2.066
20.0	2.393	2.394	2.266	2.266	2.156	2.156
30.0	2.545	2.546	2.402	2.402	2.262	2.262
40.0	2.668	2.669	2.529	2.529	2.376	2.376
50.0	2.766	2.765	2.640	2.640	2.489	2.490
60.0	2.841	2.838	2.733	2.733	2.596	2.596
70.0	2.895	2.892	2.809	2.809	2.690	2.690
80.0	2.934	2.931	2.867	2.867	2.770	2.770
90.0	2.960	2.957	2.911	2.911	2.835	2.835
100.0	2.976	2.974	2.942	2.942	2.885	2.886
110.0	2.987	2.985	2.964	2.964	2.923	2.923
120.0	2.993	2.992	2.978	2.978	2.950	2.950
130.0	2.996	2.996	2.987	2.988	2.969	2.969
140.0	2.998	2.998	2.993	2.993	2.981	2.981
150.0	2.999	2.999	2.996	2.996	2.989	2.989
160.0	2.999	2.999	2.998	2.998	2.994	2.994
170.0	2.999	3.000	2.999	2.999	2.997	2.997
180.0	2.999	3.000	3.000	3.000	2.998	2.998
190.0	3.000	3.000	3.000	3.000	2.999	2.999
200.0	3.000	3.000	3.000	3.000	3.000	3.000

Table 8: Comparison of water table heights for t = 5 days as predicted by Hybrid Finite Analytic Solution (HFAS) and Finite Element Solution (FES) for horizontal /sloping discharging aquifer with constant replenishment equal to 5 mm/day

X (m)	0 % slope		5 % slope		10 % slope	
	HFAS	FES	HFAS	FES	HFAS	FES
0.0	2.000	2.000	2.000	2.000	2.000	2.000
10.0	2.246	2.246	2.155	2.156	2.087	2.088
20.0	2.447	2.448	2.312	2.313	2.193	2.195
30.0	2.612	2.613	2.461	2.464	2.313	2.316
40.0	2.745	2.744	2.597	2.600	2.437	2.441
50.0	2.849	2.846	2.715	2.717	2.558	2.562
60.0	2.927	2.923	2.812	2.815	2.670	2.674
70.0	2.984	2.979	2.891	2.894	2.768	2.772
80.0	3.024	3.019	2.952	2.954	2.851	2.855
90.0	3.050	3.047	2.998	3.000	2.918	2.922
100.0	3.067	3.065	3.031	3.032	2.971	2.974
110.0	3.078	3.077	3.054	3.055	3.010	3.013
120.0	3.085	3.084	3.069	3.070	3.039	3.041
130.0	3.088	3.088	3.079	3.079	3.059	3.060
140.0	3.090	3.090	3.085	3.085	3.072	3.073
150.0	3.091	3.091	3.088	3.089	3.081	3.081
160.0	3.092	3.092	3.090	3.090	3.086	3.086
170.0	3.092	3.092	3.091	3.092	3.089	3.089
180.0	3.092	3.092	3.092	3.092	3.091	3.091
190.0	3.092	3.092	3.092	3.092	3.092	3.092
200.0	3.092	3.092	3.092	3.092	3.092	3.092

and 8 for t = 5 days. It can be seen from these Tables that in case of discharging aquifer also water table elevations obtained from both the solutions are almost similar. Finally the water table elevations predicted by both the solutions become constant and parallel to impermeable base. It may be seen from Tables 5-8 that contrary to a recharging aquifer, in a discharging aquifer, with increase in slope of the impermeable base, the water table height obtained from both the solutions decreases. Comparison of water table heights at t = 1 day and t = 5 days without replenishment given in Table 5 and 7 for with heights constant replenishment given in Tables 6 and 8 shows that with an increase in time the water table elevation given decreases. At t = 5 days water table profile becomes parallel to impermeable base at larger distance from the stream. Effect of constant replenishment of 5 mm/day in discharging aquifer at t = 1 day and 5 days can be realized from the fact that the values of water table elevation obtained from both the solutions given in Tables 6 and 8 than that given in Tables 5 and 7.

Conclusions

A hybrid finite analytic solution of the non-linear Boussinesq equation is presented to describe water table variation in a sloping/ horizontal aquifer receiving constant replenishment and interacting with the adjoining stream having abrupt rise or drop of water level. Both hybrid finite analytic solution and finite element numerical solution give almost similar water table elevations. The water table profiles obtained from both the solutions finally attain a constant value and become parallel to impermeable base. Both the solutions show that in case of recharging aquifer, the water table is higher at t = 5 days than at t = 1 day, whereas in case of discharging aquifer, the water table is higher at t = 1 day than at t = 5 days. Due to the effect of constant replenishment with the rate of 5 mm/day, water table is higher compared to that for zero replenishment, both at t = 1 day and t = 5 days. It is suggested that hybrid finite analytic solution of the nonlinear Boussinesq equation should be preferably used due to its simplicity in computations as compared to the finite element numerical solution, which involves complicacy in mathematics and computer programming.

References

Chen, C.J., 1988. Finite analytic method. In: Handbook of numerical heat transfer, eds.: Minkowyez, W.J., Sparrow, E.M., Schneider, G.E., and Pletcher, R.H., John Wiley and Sons Inc., New York, 723-746.

Lockington, D.A., 1997. Response of unconfined aquifer to sudden change in boundary head. J. Irrig. and Drain. Engrg., 123 (1), 24-27.

Marino, M.A., 1973. Water table fluctuation in semipervious stream-unconfined aquifer systems. J. Hydrol., 19, 43-52.

Ozisik, M.N., 1980. Heat conduction. John wiley & sons, U.S.A.

Pi, Z., and Hjelmfelt, A.J., 1994. Hybrid finite analytic solution of lateral subsurface flow, Water Resour. Res., 30(5), 1471-1478.

Polubarinova-Kochina, P.Ya., 1962. Theory of ground water movement. Princeton University Press, Princeton, N.J.

Serrano, S.E., and Workman, S.R., 1998. Modelling transient stream/aquifer interaction with the non-linear Boussinesq equation and its analytical solution. J. Hydrol., 206, 245-255.

Sidiropoulos, E., Asce, A.M., Tzimopoulos, C., and Tolikas, P., 1984. Analytical treatment of unsteady horizontal seepage. J. Hydraulic Engrg, 110 (11), 1659-1670.

Tolikas, P.K., Sidiropoulos, E.G., and Tzimopoulos, C.D., 1984. A simple analytical solution for the Boussinesq one-dimensional ground water flow equation. Water Resource Res., 20 (1), 24-28.

Upadhyaya, A., and Chauhan, H.S., 1998. Solutions of Boussinesq equation in semi-infinite flow region. J. Irrig. and Drain. Engrg., ASCE, 124 (5), 265-270.

Upadhyaya, A., 1999. Mathematical modelling of water table fluctuations in sloping aquifers. PhD thesis, G. B. Pant University of Agriculture and Technology, Pantnagar, U.P. - 263145, India.

Upadhyaya, A. and Chauhan, H.S., 2001a. Interaction of stream and sloping aquifer receiving contant recharge. J. Irrig. and Drain. Engrg., ASCE (In Press).

Upadhyaya, A. and Chauhan, H.S., 2001b. Falling water tables in horizontal/sloping aquifers. J. Irrig. and Drain. Engrg., ASCE (In Press).

Workman, S.R., Serrano, S.E., and Liberty, K., 1997. Development and application of an analytical model of stream/aquifer interaction. J. Hydrol., 200, 149-164.

Yussuff, S.M.H., Chauhan, H.S., Kumar, M. and Srivastava, V.K., 1994. Transient canal seepage to sloping aquifers. J. Irrig. and Drain. Engrg., 120 (1), 97-109.

Hydrogeological characterization of fractured hard rock aqufers: Assessment of modelling approaches

M. Razack and O. Gaillard

Department of Hydrogeology UMR 6532, University of Poitiers, 40 avenue du Recteur Pineau, 86022 POITIERS Cedex, France, e-mail: moumtaz.razack@hydrasa.univ-poitiers.fr

Abstract

The Hydrogeology of fractured massive rocks still suffers from a lack of in situ research facilities, whereas to date the advances in modelling methods (analytical and/or numerical) of these aquifers are very significant. The choice between several approaches of modelling is obviously related to the comprehension of the medium. The need for carrying out precise in situ experiments and observations and for characterising the medium thoroughly is recognised without ambiguity by the community of hydrogeologists (Long et al., 1996). The installation of equipment of in situ research in fractured medium must make it possible to collect reliable data in quantity and quality concerning the site (flow, transport, fracture geometry, water-rock interactions... data) and to evaluate the tools and methods for characterising these media.

An experimental site was thus set up in fractured massive rocks (site of Gandouin, France) constituted of Silurian quartzitic sandstones. Several models of flow in fractured rocks were developed during the last decades (linear, spherical, fractal models). We propose in this paper to examine the applicability of these approaches by holding account in particular of the degree of compartmentation of the tested medium. Several pumping tests were conducted at this site and permitted to gather reliable hydraulic data, on which classical and more recent models could be tested. We show thus that in this type of medium the applicability of the continuum model is related on investigation scale. The contribution, in term of dimension of flow, of recent approaches based on fractal methods is shown when the continuum model is not applicable any more.

Key words Fracture aquifers, pumping tests, analytical models, flow dimension

Introduction

Hydrogeology of fractured massive rocks still suffers from a lack of in situ research facility, whereas to date the advances in modeling methods (analytical and/or numerical) of these aquifers are very significant. The choice between several approaches of modeling is obviously related to the comprehension of the medium. The community of hydrogeologists (Long et al., 1996; Allen and Michel, 1997) recognizes without ambiguity the need for carrying out precise in situ experiments and observations and for characterizing the medium thoroughly. The complex hydraulic behavior often observed within fractured hard rocks aquifers, resulting from the major fractures orientations and the rocks matrix permeability, has in general been often described in qualitative and semi-quantitative way (Smith and Vaughan, 1985). The installation of equipment for in situ research in fractured medium must make it possible to gather reliable data in quantity and quality concerning the site (flow, transport, geometry, water-rock interactions...) and to evaluate the tools and methods for characterizing these media. Accordingly an experimental site was set up in fractured massive rocks (site of Gandouin, Morbihan, France) (Gaillard, 2001).

Several flow models in fractured medium were developed during these last three decades (Thiory et al., 1983; Gringarten & Witherspoon, 1972). Recently new analytical approaches were proposed, based on fractal models (Barker, 1988). We propose in this paper to examine the applicability of these approaches by holding account in particular of the degree of compartmentation of the tested medium.

The experimental site

The experimental site of Gandouin (Fig.1) is located at the southeast of the town of Malestroit in the department of Morbihan (Bretagne, Western France). The site is an old pit, which was exploited from 1978 to 1990 for aggregate extraction. This pit is not any more in exploitation from now on and is used as storage for sterile detrital from the Roga pit located a few kilometres to the South. The stop of the exploitation involved the formation of a water body in place of the cavity because of the mine pumps stopping.

The experimental site occupies a surface of approximately 12 hectares. The water level currently has a surface of 2.5 ha and a volume estimated about 375 000 mV. The water level is relatively deep-cut and surrounded by the old front , which has an average height of twenty meters above the water level.

The water level is accessible only by one approach ramp which was arranged in the western part of the site allowing the setting of a boat to carry out the various surveys (mainly fracturing surveys) necessary to this study.

Figure 1C represents a simplified geological map of the site of Gandouin. The site is located on the Malestroit synclinal. The formation of Gandouin, made of pyritous quartzic sandstones, is surrounded on both sides by siltstones (to the North by the formation of Saint-Marcel and to the South by the Bois-Neuf formation) and is posed on St-Marcel siltstones formation.

In the Southern part of the site one distinguishes the Traveusot siltstones and a formation of Armorican sandstones which continues to the granites of Lanvaux. The northern and southern parts of the site are delimited by a major regional fault striking N110 (fault of Malestroit). Finally on the eastern part one distinguishes a sandy formation of Pliocene age.

The Gandouin sandstones are laid out in strongly tilted layers and present a dip in the order of 70 degrees to the South (Corre, 1969).

The formation of Gandouin is assimilated to lower Llandovery (beginning of Silurian). It is constituted of pyritous quartzic sandstones. This formation presents an average thickness of 50 meters. These quartzic sandstones appear of beige colour, rather coarse and of saccharoid aspect. They also contain intercalations of black and pyritous siltstones. The pyrites formation is due to the presence of a reducing medium resulting from a low water thickness during the diagenesis of the Silurian formations. These intercalations can be laid out in the form of very local deposits. The presence of these schists posed many technical problems at the time of the boreholes drilling. One finds at various places of the site these deposits of black schists and in particular in the Western zone of the site.

The site of Gandouin presents a well-developed fractures network. The main structure of this area is due to the Hercynian tectogenesis. Faulting tectonics is very significant. The major accidents are easily identifiable on the sandstone formations. The southern side of Malestroit synclinal is delimited by the Malestroit fault. The Saint Marcel fault separates the synclinal northern side where the layers are almost vertical from the central part in which the Silurian sandstones (to which the formation of Gandouin belongs) draw an anticline sometimes disymetric and slightly poured towards North. The major fault corresponds to the Malestroit fault.

A series of fracture measurements was carried out using a geological compass in April 1998 on the various pit fronts. A hundred measurements could have been be carried out. The measured parameters were the orientation, the dip, the extension, the opening of the fractures and the linear fracturing density. The fracture orientation diagram (Fig.1B) points out two major orientations : a first fractures set of orientation N0 and a second one of orientation N110. These two orientations correspond to the regional major directions. The for-

114

mation of Gandouin presents a general dip of 70 degrees towards the south. Measurements of extension were carried out directly on the sandstone outcrop. The average extension of the fractures is in the order of ten meters. The linear density of fracturing was also measured on the various fronts and exposures. This density corresponds to the number of fractures per meter in a given direction. The densities could be measured (in fractures per meter) for the two principal sets of fractures met on the site. One obtains a density of 3 fractures/meter for the set striking N0 and 6 fractures/meter for the set striking N110. Thus fracturing is denser in the N110 direction, which is the principal set. This direction also corresponds to the principal flow direction. The study of the fracturing of the experimental site played a role in the choice of the boreholes layout in the field. Indeed, boreholes were placed according to the principal fracturing directions, which are N110 and N0. This study constitutes a valuable and essential help for interpreting the pumping tests carried out on the experimental site because the flow in fractured media is in direct relationship with the fracturing trends.

The site is equipped with 6 boreholes (PZ1 to PZ6) going from 15 to 40 meters depth (Figure 1A). One distinguishes two zones on the site with quite distinct characteristics. Boreholes of the Eastern zone (boreholes PZ1, PZ2, PZ3) intersected a relatively continuous medium made up of 10 meters filling, then of approximately 30 meters of fractured quartzic sandstones. These three borcholes thus penetrate almost completely the sandstone aquifer in the eastern zone of the site.

Conversely, boreholes of the Western zone of the site crossed a much more compartmented medium showing intercalations of fractured quartzic sandstone and less permeable black schists. This compartmentation is better marked on boreholes PZ4 and PZ6 than on PZ5. This indicate how complex is the hydrodynamic behaviour of the sandstone aquifer.

Hydraulic tests and interpretation

Two hydraulic experiments were carried out on this site (September 98 and March 99). The test schedule is detailed in the table-1. It should be noted that all boreholes of the East zone reacted well between them to the various pumpings carried out on this zone. Which confirms the apparent continuity of the medium here. On the other hand no reciprocal influence was noted clearly between borehole of the Western part. This gives an indication of the complexity and the compartmentation of this zone.

The behaviors observed on the two zones of the site are very different (Figure 2). The Eastern zone of the site in which boreholes PZ1, PZ2 and PZ3 intersected fractured sandstones, acts as an EPM (Equivalent Porous Medium) at this macroscopic scale of analysis (Fig.2(A)). The drawdown-time graph emphasises a radial flow towards the pumped well, obeying the continuum model of Theis. Let us note on this graph, the very marked effect of the boundary represented by the water body. On the other hand the hydraulic behavior of the Western zone of the site (Fig.2(B)) is much more complex and is absolutely not connected with an EPM behavior. The data collected made it possible to test, compare and assess the validity of the theoretical models suggested in the literature. As a first simplification, the fractured medium can be assimilated to an equivalent porous medium (EPM). The fractured medium can also be represented by a model with a single fracture. Finally the concept of flow dimension can be integrated through a fractal approach.

Accordingly we tested the universal continuum Theis model, the single fracture Gringarten's model and the flow dimension Barker's model. These models are briefly presented hereunder. Note that the permeability of the quartzic sandstones is very low thus we considered that flow takes place only through the fractures and not in the matrix. Thus models based on the concept of double porosity (Barenblatt, 1960 ; Moench, 1984) are not considered here.

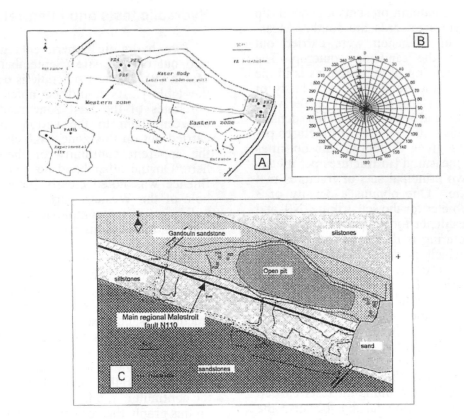

Figure 1: Experimental site of Gandouin (France). A: Site map showing location of the boreholes; B: fracture orientation diagram ; C : simplified geological map of the site.

Table-1: Pumping tests schedule at the Gandouin site

Period	Pumping well	Duration	Discharge m3/h	observation
Sept 98	PZ1	32 h	12.4	PZ1, PZ2, PZ3
	PZ2	58 h	12	PZ1, PZ2, PZ3
	PZ4	72 h	13	PZ5, PZ6
April 99	PZ5	24 h	0.8	PZ4, PZ6
	PZ6	12 min	0.6	PZ5, PZ6

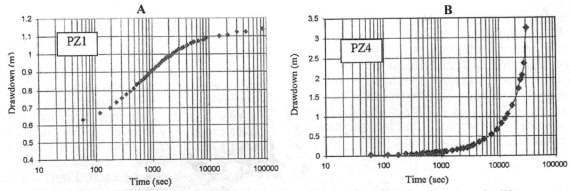

Figure 2: Drawdown vs. time curve observed on the Eastern zone (borehole PZ1) and on the Western zone (borehole PZ4) of the site

Theis model

The model of Theis (1935) is well known and routinely used in hydrogeology. The model requires a continuous medium. The applicability of this model implies that at the considered investigation scale, the continuity assumption is verified. The Theis model is expressed as follows:

$$s = \frac{Q}{4\pi T}[W(u)]\dots\dots\dots\dots\dots[1]$$

where s = drawdown (m), Q = discharge (m^3/s), T= transmissivity (m^2/s), S = storage coefficient (-), t = time (s), r = distance between the pumped well and the observation well (m), $W(u)$: Well function, $u = r^2S/4Tt$. Approximations of $W(u)$ can be find in (bramowitz, (1970).

In the presence of a water body, using the superposition principle, the solution becomes:

$$s = \frac{Q}{4\pi T}[W(u) - W(u')]\dots\dots\dots[2]$$

where r'=distance from the image well to the observation well, $u'=r'^2S/4Tt$.

Gringarten's model

The model of Gringarten (Gringarten et al., 1972) is a model with one single fracture and considers the pumping well to be intersected by a single fracture that is significantly more transmissive than the rest of the aquifer. This model characterises that fracture based on drawdown data from the pumping well that typically plots a straight line on a log-log diagram at early time. The log-log curve merges with a Theis curve if the pumping test is sufficiently long. The flow is thus linear in an elliptical zone close to the fracture. Beyond this elliptical zone of linear flow, drawdown in observation wells may be interpreted with Theis model (radial flow) for large distances or after long period of pumping. The single fracture is referred to

in the literature as an extended well (Jenkins & Prentice, 1982). Figure 3 shows the case of a single vertical fracture. At the pumping well and at early pumping times (beginning of pumping), Gringarten's solution is written as follows:

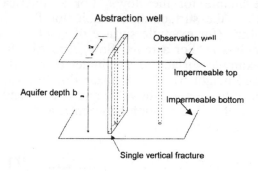

Figure 3: Single fracture flow model

$$s = \frac{Q}{4\pi T_f}\left[\sqrt{\frac{\pi}{u}}\,\text{erf}\left(\sqrt{u}\right) + W(u)\right]\dots\dots[3]$$

$$u = \frac{x_f^{\,2}S}{4T_f t} \dots\dots\dots\dots\dots[4]$$

where T_f = fracture transmissivity [L^{2}T^{-1}], x_f = fracture half-length [L], erf(x)= error function, $W(u)$ well function. At the pumping well and for late times, the model expression is:

$$s = \frac{Q\sqrt{t}}{2x_f\sqrt{\pi TS}} \dots\dots\dots\dots\dots[5]$$

The Gringarten model permits to obtain the parameter xf, which corresponds to the half-length of the fracture. This parameter can be useful to have an idea of the geometry of the fractured medium and in particular the level of its extension.

Barker's model (1988)

The concept of fractal dimension can be applied to the hydrodynamics of fractured media. Indeed its application to the study of the fractures networks (Bodin & Razack, 1999) as to fluids flow within the

117

network, is based on the fact that large fractures often derive from the coalition of smaller fractures.

Barker (1988) introduced the concept of entire and non-entire flow dimension for the analysis of hydraulic test. The dimension of the flow describes the relation between the test distance and the surface A available for the flow. For a spherical flow the surface A available for flow is $4\pi r^2$ (for example surface of a sphere of radius r). For a radial flow, $A = 2\pi r b$ (for example surface of a cylinder of radius r and height b). For a linear flow the surface has a constant value and is independent of r. The relation between A and r can be generalised by:

$$A = \alpha_D r^{D-1} \dots\dots\dots\dots\dots [6]$$

where D = entire euclidian dimension (D=1, 2 or 3), α_D is a constant of proportionality which is related to D.

What is the contribution of Barker's solution compared to the other methods? The assumption of a fractal dimension improves the flow analysis in the fractured aquifers. Indeed, it is not limited to the unique fracture intersected by the pumping well, as in the case of Gringarten's model with single fracture, but takes into account, even in a synthetic way, spatial arrangement of the fractures as a whole.

Figure 4 compares flows in continuous porous medium and fractured medium. This figure also visualises the form of the flow according to its dimension. A non-entire dimension of the flow represents hybrid flow geometry. For example, a dimension ranging between 2 and 3 characterises flow geometry between radial and spherical. In the same way a dimension ranging between 1 and 2 represents hybrid geometry of flow between a linear and radial flow.

Barker assumes that the dimension of the flow can be related to the fractal dimension of the fracture network in which the flow takes place. The flow dimension can be lower or equal to the fracture network dimension. Barker's solution is expressed as follows:

$$h(r, t) = \frac{Qr^{2N}}{4\pi^{1-N}K_f b^{3-D}} \Gamma(-N, u) \dots\dots [7]$$

$$N<1 \qquad u = \frac{S_{sf}r^2}{4K_f t} \qquad \dots\dots\dots [8]$$

where h(r,t) = hydraulic head [L] at time t and at a distance r from the source, r is the radius [L]; $\Gamma(a, x) =$ incomplete Gamma function (see Abramowitz, 1970); D is the flow dimension [-]; N = 1-D/2 ; K_f = hydraulic conductivity [LT^{-1}] ; Q = discharge [$L^3 T^{-1}$]; S_{sf} = specific storage [L^{-1}] ; b is a coefficient which expresses that, if the flow has an unspecified dimension , it remains included in the Euclidean space of dimension three.

By using particular values of the Gamma function one obtains for dimensions 1, 2 and 3 the following functions:

$$D=1 \Rightarrow h(r,t) =$$
$$\frac{Qr}{4\sqrt{\pi}K_f b^2}\left(\frac{\exp(-u)}{\sqrt{u}} - \sqrt{\pi}erfc\left(\sqrt{u}\right)\right)$$
$$\dots\dots\dots\dots\dots\dots\dots [9]$$

$$D=2 \Rightarrow \quad h(r,t) = \frac{Q}{4\pi K_f b}W(u)\dots\dots[10]$$

$$D=3 \Rightarrow \quad h(r,t) = \frac{Q}{4\pi K_f b}erfc\left(\sqrt{u}\right) \dots [11]$$

Equations [7] to [11] reveal that for large times and for D<2, the time-dependant term is dominating in Barker's solutions. When D>2 the time-dependant term tends to zero. Hence a steady flow during pumping will be achieved when and only the flow dimension is greater than two.

One can note that expression [10] corresponding to Barker's solution for D=2 is equivalent to Theis' solution [1]. Barker's solution can thus be taken as a generalisation of Theis solution.

When D=3, the flow is spherical, and all the fractures are solicited by the flows.

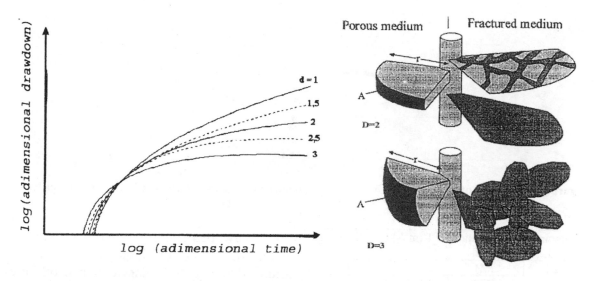

Figure 4: Barker's model type curves and corresping flow in porous and fractured media

When D=2, the flow is radial, and intersect only certain fracture planes. When D=1, the flow is linear and takes place in no more than one or two channelled fractures. The dimension of the flow can also be non-entire when it presents an intermediate form. Standard curves of drawdown vs. time according to Barker's solution are shown in figure 4B. Pumping tests interpretation and results

All calculations of pumping tests data adjustment were made with the use of the software EXCEL™ and the optimisation functions provided with the software.

Interpretation according to Theis model

The time-drawdown curves obtained for the eastern zone pumping wells and observation wells (PZ1, PZ2, and PZ3) and for well PZ5 (Western zone) could be interpreted according to this model (Fig.5). Hence, on a decametric scale, the fractured sandstones can be regarded as continuous. This assumption is made possible as the fracturing of the sandstones is relatively intense and seems to present a rather good connectivity. One should note, however, that Theis model fitted to the observed data suggests that early drawdowns in these wells are slightly more than predicted by the model.

On the other hand, pumping test data of wells PZ4 and PZ6 (Western zone of the site) could not be interpreted with Theis solution because of the compartmentation of the aquifer in this zone.

Accordingly, the eastern zone of the site could be characterised hydraulically, assuming an equivalent continuum at a decametric scale. Results are reported in Table 2. Transmissivity ranges from 2.10^{-3} m²/s to 8.10^{-3} m²/s and storage coefficient ranges from 1.10^{-3} to 5.10^{-2}. Transmissivity at borehole PZ5 on the Western zone is much lower, i.e $T= 4.10^{-5}$ m²/s. We should be aware, that though fitting between measured data and the Theis model is rather good, such a characterisation remains just global and does not provide any precise information about the aquifer.

Note that the wellbore storage is not taken into account in these analyses. As the boreholes diameter is small (115mm) and no gravel was placed, we assumed that the borehole storage effect could be negligible in a first approach. Further analyses are projected using the derivative method suggested by Spane and Wurstner (1993).

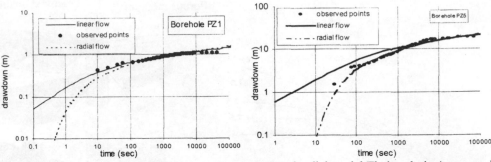

Figure 5: Adjustment of linear model (Gringarten's solution) and radial model Theis solution) to pumping tests data.

Pumping well	Observation well	T (m²/s)	S (-)
PZ1	PZ1	2.3 10⁻³	4.9 10⁻²
	PZ2	3.1 10⁻³	1.3 10⁻²
	PZ3	6.2 10⁻³	1.9 10⁻²
PZ2	PZ1	3.8 10⁻³	1.1 10⁻²
	PZ2	7.8 10⁻³	1.3 10⁻²
	PZ3	8.1 10⁻³	2.9 10⁻²
PZ5	PZ5	4.2 10⁻⁵	6.2 10⁻³

Table-2: Pumping tests interpretation according to Theis model (radial flow)

Interpretation according to Gringarten's model

Fitting Gringarten's model to pumping test data requires 3 parameters which are i) the fracture transmissivity (T_f), ii) the storage coefficient (S) and iii) the fracture extension ($2x_f$). When the fracture extension is not known, and one looks only for a numerical adjustment between observed and computed data, then the problem is undetermined (several combinations of these 3 parameters exist that yield quite good adjustment). Gringarten's model requires preliminary knowledge of the fracture length. The measures, which we took on the site, resulted with an average fracture length value of 10m. We thus introduced this value into the model of Gringarten. The values of T and S were then estimated by optimisation (best fitting between observed and computed data). Note that wells with single fracture flow typically present drawdown curves which are straight lines (linear flow) with slopes of 0.25 to 0.5 (Streltsova, 1988). A slope of 0.25 is related to a low permeability fracture, whereas a slope of 0.5 signifies high permeability fracture. Pumping tests data obtained on boreholes PZ1, PZ2 (Eastern block) and PZ5 (Western block) could be interpreted using Gringarten's model. Fittings are shown in figure-5 and results are reported in Table 3.

Table 3: Pumping tests results for wells PZ1, PZ2 and PZ5 according to Gringarten's model (linear flow at early times)

Pumping well	T_f (m²/s)	S	Fracture length
PZ1	2.4E-03	5.8 E-04	10 m
PZ2	5.7E-03	9.6 E-05	10 m
PZ5	8.9 E-05	7.4 E-04	10 m

Single fracture model yields fracture transmissivity values comparable to the EPM (equivalent porous medium, radial flow) transmissivity. Single fracture model storage coefficients are on the other hand significantly lower. These interpretations show that a linear flow, if ever it occurs, would be very short (about a few tens of seconds). The flow tends rather quickly towards a radial (or pseudo-radial) form. The influence of pumping is sufficiently extended at that time to intersect a sufficient number of fractures and to take on a radial geometry.

Interpretation according to Barker's model with entire dimension

Solution for 1D was not used because it might not be representative of what would occur in the natural fracture network at the Gandouin site (see preceding section on Gringarten's model).

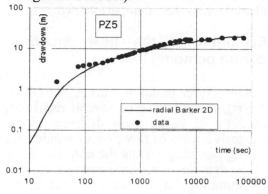

Figure 6: Barker's 2D model fitted on PZ5 pumping data

Table 4: Main hydrodynamic characteristics obtained using Barker's 2D method. K_f is the fracture hydraulic conductivity, S_s the specific storage, b is the thickness of the aquifer.

Pumping well	Obs. well	K_f (m/s)	S_s (m^{-1})	b (m)
PZ5	PZ5	5.10^{-7}	5.10^{-3}	15
PZ1	PZ1	$1,36.10^{-4}$	$1,37.10^{-3}$	25
	PZ2	$2,48.10^{-4}$	$2,25.10^{-4}$	25
	PZ3	$5,1.10^{-4}$	$3,58.10^{-4}$	25
PZ2	PZ1	$4,23.10^{-4}$	$4,63.10^{-5}$	25
	PZ2	$3,61.10^{-4}$	$3,47.10^{-4}$	25
	PZ3	$8,5.10^{-4}$	$8,5.10^{-4}$	25

Solution 2D is equivalent to Theis solution (radial flow). Hence rather close results were obtained according to these two methods for pumpings on PZ1, PZ2, PZ3 (East side) and PZ5 (West side) (Table 4 and Figure 6). One should note however, that Barker's 2D solution is written with K_f (fracture hydraulic conductivity) and S_s (specific storage coefficient) and thus requires previous knowledge of the aquifer's thickness (b). If one assumes that the flow can be spherical, it is possible in this case to use Barker's 3D model. As an example two adjustments are proposed concerning pumping PZ5 (Western zone) and pumping PZ2 with observation on PZ1. The results are given in table 5 and figure 7.

Table 5: Main hydrodynamic characteristics obtained using Barker's 3D method. K_f is the fracture hydraulic conductivity, S_s the specific storage, b is the thickness of the aquifer.

Pumping well	Obs. well	K_f (m/s)	S_s (m^{-1})	b (m)
PZ5	PZ5	1.10^{-7}	5.10^{-4}	15
PZ2	PZ1	$4.5\ 10^{-5}$	$2.4\ 10^{-4}$	25

A spherical geometry supposes that all the fractures are requested by the flow. The graphs show that the adjustment of the experimental data to the 3D model is much better for large times of pumping, than at the beginning of pumping. On the Eastern zone, this fact can be attributed to the effect of the recharge boundary (water body). On the Western zone (PZ5), the long-term evolution of pumping at PZ5 cannot clearly be attributed to the water body because the geological cross sections do not show continuity of the sandstones between the boreholes and the water body. It is thus probable that the flow takes in this context in effective spherical form. It should be noted that the 2 tested borholes, the permeability of fractures is rather low (hydraulic conductivity K_f ranging between 10^{-7} m/s and 10^{-5} m/s).

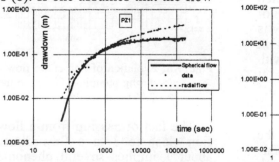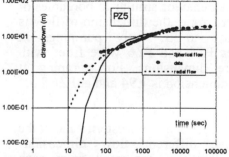

Figure 7: Barker's 3D model fitted to pumping test data on boreholes PZ1 and PZ5

121

Table 6 : Interpretation of pumpings PZ4 and PZ6 with Barker's method. Evolution of the flow dimension during pumping. KF is the hydraulic conductivity of fracture, Ss the specific storage, D the flow dimension.

Pumping	Time Interval	K_f (m.s^{-1})	S_s (m^{-1})	D (-)
PZ4	60 ;1200 s	$8,43.10^{-2}$	5.10^2	2,67
	1200 ; 7200 s	$7,63.10^{-1}$	1.10^2	2,22
	7200 ; 35280 s	$2,41.10^{-1}$	1.10^2	1,798
PZ6	38 ; 160 s	$1,14.10^{-1}$	$4,32.10^{-3}$	2,855
	160 ; 600 s	1.10^{-3}	$2,9.10^{-2}$	2
	600 ; 720 s	$1,26.10^{-4}$	$1,13.10^{-2}$	1

Pumpings at on PZ4 and PZ6 could not be interpreted with Barker's model with entire flow dimension as not any satisfactory adjustment could be produced.

Barker's model with non-entire flow dimension

Adjustments with non-entire flow dimensions were carried out in order to understand the various forms, which the flow in the medium could take. Equation [7] was used to calculate the theoretical drawdown vs. time curve, which then was compared with the observed drawdown curve. In addition to parameters K_f, S_f and b, the parameter D corresponding to the non-entire flow dimension can be adjusted.

For pumpings of the eastern zone, a dimension close to 3 still yields the best adjustment. For pumping at PZ5, which presents a behavior similar to boreholes of the eastern zone, the dimension of flow is still close to 3. Only pumpings at boreholes PZ4 and PZ6 were fitted with non-entire dimensions. For pumping at PZ4 the dimension is 1.94 and at PZ6 2.5.

One can see that boreholes PZ4, PZ5 and PZ6, located in the western zone of the site, have different flow dimensions whereas they are separated from each other by only ten meters. It suggests again that the medium in the zone of the experimental site is very compartmented

Evolution of the flow dimension during pumping test

On certain pumping curves log (s) vs. log (t), the presence of several rectilinear portions of curve with different slopes suggests a form of flow, which would vary during pumping. Thus the concave shape of the drawdown vs. time curves measured for pumpings at the western zone of the site (pumping PZ4 and PZ6) can be interpreted in term of variation of flow and hence of evolution of flow dimension during pumping. The determination of dimension is made on each rectilinear portion of the curve. Adjustment of parameters K_f, S_s and D is carried out independently for each portion of curve.

In case of pumping at PZ4, an adjustment of parameters K_f, S_s and D was carried out while distinguishing parts on the observed curve 3 (Figure 8). One can notice that the dimension of flow evolves successively from 2.67 to 2.22 then to 1.79. The same type of adjustment of the flow parameters was carried out for the pumping at PZ6. The results of the various parameters obtained for pumpings at PZ4 and PZ6, are summarised in table 6 (K_f: hydraulic conductivity of fracture, S_s: specific storage, D: non-entire flow dimension).

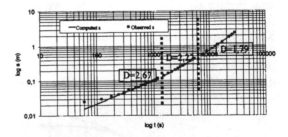

Figure 8: Evolution of non entire flow dimension during pumping. Fitting to the borehole PZ4 pumping data

The fact of passing from a flow dimension close to 3 to a flow dimension of about 1 implies several phenomena. At the beginning of pumping the flow is rather spherical then tends to become radial and finally evolves towards a linear form. It is possible to interpret this phenomenon by the fact that at the beginning of pumping, the aquifer is solicited in its entirety and in all the space directions (spherical flow). Then progressively with pumping, the aquifer is solicited only on certain fractures (radial flow) because the aquifer is compartmented. Between the spherical flow and the radial flow, connectivity between the fractures would probably decrease. At the end of the pumping, there remain only some fractures (even a single fracture, or part of fracture, when it is channelled) which take part in the flow (linear flow). We tried also to analyse pumpings of the eastern zone (PZ1 and PZ2) with this same approach. The results, compared to the Western zone of the site, are very different. One can see an increase (though very rapid) of the dimension of flow. The results (Table 7) show that, contrary to the evolution observed on the Western zone, the flow evolves very quickly towards a spherical form ($D\approx3$). This form of flow in fact is caused by the presence of the water body acting like a recharge boundary. All the fractures thus remain solicited by the flow.

Table 7. Interpretation of pumpings at PZ4 and PZ6 with Barker's method. Evolution of the flow dimension during pumping. KF is the hydraulic conductivity of fracture, Ss the specific storage, D the flow dimension.

pumping	Observation well	Time Interval(s)	K_f (m.s^{-1})	S_s (m^{-1})	D (-)
PZ1	PZ2	60 - 900	$1.2*10^{-3}$	$2.65*10^{-4}$	2,52
		960 - 2280	$8*10^{-4}$	$1.02*10^{-4}$	3
	PZ3	60 - 900	$1.96*10^{-4}$	$6.36*10^{-4}$	3
		960 - 2280	$1.96*10^{-4}$	$6.36*10^{-4}$	3
PZ2	PZ1	60 - 900	$1.02*10^{-3}$	$2.96*10^{-4}$	2,52
		960 - 2400	$8*10^{-4}$	$1.02*10^{-4}$	3
	PZ3	60 - 900	$5*10^{-3}$	$8*10^{-3}$	1
		960 - 2400	$1.37*10^{-2}$	$1.61*10^{-4}$	3

Barker's method with the use of non-entire flow dimension allows interpreting certain pumping tests in case of a complex compartmented medium as is the Western zone of the site (boreholes PZ4 and PZ6). The use of non-entire dimensions also allows characterising the form of the flow within the fractured aquifer.

However the physical interpretation of the non-entire flow dimension is the most difficult conceptual feature of Barker's model and is not clear. It might be expected that flow non-entire dimension would be related to some characteristic statistical property of the fracture network. According to Bangoy et al., (1992), during a pumping with constant discharge, flow dimensions $D\geq2$ lead to a steady regime at late pumping times. On the other hand, flow dimensions $D\leq2$ give, at each unsteady stage of linear flow, on a log (s) vs. log (t) diagram, a line with a slope $N = 1-D/2$ where D is an apparent dimension of flow for a given investigation distance. This flow dimension, which evolves with time, from a unidimensional flow (D=1) to a radial or pseudo-radial flow (D=2), would depend on contrast between hydraulic conductivities of the fracture and the rock matrix.

Conclusion

The experimental site of Gandouin (Morbihan, France) in the region of fractured rocks made it possible to gather reliable hydraulic data, on which several flow models (linear model, radial model, spherical model, non entire fractal dimension model) could be tested. Fracture analysis revealed 2 major fracture sets (N0 and N110) and a highly fractured site, suggesting that radial flow might develop through interconnected fractures. Pumping tests revealed however two very different zones on the site: an East zone behaving like an equivalent porous medium; a West zone compartmented and displaying a complex bahaviour.

On the East zone, it was found that the fractured aquifer responds quite well with a radial model. Occurrence of linear flow at early pumping times is probable but not clearly pointed out through the tests.

The contribution, in term of flow dimension and comprehension of the flow form and its evolution with time, of recent approaches based on fractal description of the medium is shown when the continuum model is not applicable any more. The introduction of the dimension concept in flow models is undoubtedly an unquestionable progress although difficulties persist in its interpretation. This flow dimension would be related to some statistical property of the fracture network. We showed however that this approach permitted to characterise flow in a complex compartmented medium.

Analysing pumping tests data in fractured hard rocks aquifers requires in principle some knowledge about the geometry of the fracture system. However as there is no analytical method which could be universally applied to such pumping tests data, the usefulness of applying several approaches is here demonstrated.

References

Abramowitz, M. and Stegun, I.A., 1970. Handbook of mathematical functions with formulas, graphs and mathematical tables. Dover Publications, Inc, New York, 1046 pp.

Allen D.A. and Michel F.A. (1998). Evaluation of multi-well test data in a faulted aquifer using linear and radial flow models. Ground Water, 36(2) :938-948.

Bangoy, L.M., 1992. Hydrodynamique d'un site expérimental en aquifère du socle fissuré; nouvelle méthode d'interprétation des essais hydrauliques. Thèse de doctorat Université de Montpellier (France), 138 pp.

Barenblatt, G.E., Zheltov, I.A. and Kochina, I.N., 1960. Basic concepts in the theory of seepage of homogeneous liquids in fissured rocks. Appl. Math Mech., 24: 1286-1303.

Barker, J.A. (1988). A generalised radial flow model for hydraulic tests in fractured rock. Water Resour. Res., 24(10): 1796-1804.

Bodin J. and Razack M. (1999). A method based on image processing for the analysis of fractures networks geometry. Geometrical properties and scaling laws. Bull.Soc.Géol.Fr., 170(4) :579-593.

Corre, C.L., 1969. Contribution à l'étude géologique des synclinaux du sud de Rennes (Massif armoricain). Thèse 3ème cycle Thesis, Orsay, 116 pp.

Gringarten, A.C. and Witherspoon, P.A. (1972). A method of analyzing pump test data from fractured aquifers. In: I.A.o.R. Mechan. (Editor), Percolation through fissured rock. Deutsche Gesellschaft für Red and Grundbau, Stuttgart (deutschland), pp. T3B1-T3B8.

Jenkins D.N. and Prentice J.K. (1982). Theory for aquifer test anlysis in fractured rocks under linear (non radial) conditions (Ground Water, 20(1): 12-21.

Long J.C.S., A. Aydin, S.R.Brown et al., (1996). Rock fractures and fluid flow : comtemporary understanding and applications. Nat.Acad.Press, Washington D.C., 560 p.

Moench, A.F., 1984. Double-porosity models for a fissured groundwater reservoir with fracture skin. Water Resources Research, 20(7): 831-846.

Smith E.D. & Vaughan N.D. (1985). Aquifer test analysis in nonradial flow regimes : a case study. Ground Water, 23(2) :167-175.

Spane F.A. & Wurstner S.K. (1993). DERIV : a computer program for calculating pressure derivatives for use in hydraulic test analysis. Ground Water, 24(5) :814-822 ;

Theis, C.V. (1935). The relation between the lowering of the piezometric surface and the duration of discharge of a well using groundwater storage. Amer. Geophys. Union, Transactions, 16: 519-524.

Thiery D., Vandenbeusch M. et Vaubourg P.(1983). Interprétation des pompages d'essai en milieu fissuré aquifère. Doc BRGM., 57, 53p.

Intl. Conf. on Sustainable Development and Management of Groundwater Resources
in Semi-Arid Region with Special Reference to Hard Rock, (IGC 2002),
M. Thangarajan, S.N. Rai & V.S. Singh (Eds.)

A numerical model for simulating transport and geochemistry in groundwater

K.V.Hayagreeva Rao and M.Sekhar
Indian Institute of Science, **Banglore** – 560 012, **India,**
e-mail: kvhrao@civil.iisc.ernet.in

Abstract

In this paper, a model is presented which combines the transport with the equilibrium aqueous geochemistry for simulating hydrogeochemical behavior observed in groundwater system. The model is capable of considering ion-exchange, precipitation-dissolution, redox and acid-base reactions occurring in groundwater. The modeling framework uses a coupling between transport module with popular equilibrium aqueous speciation module MINTEQA2 based on a sequential iterative approach which is commonly used for reactive transport. The transport is modeled using a finite volume approach in order to simulate both advection and dispersion dominated systems.

The numerical model is applied on three field problems to demonstrate the applicability of such a model for analysing groundwater system influenced by transport and geochemical reactions. Two of the above problems pertain to column experiments which are idealisations of field situations. The first problem corresponds to an ion-exchange case. The second problem pertains to the chemical concentration patterns evolving from complex changes in chemical compositions resulting from precipitation and dissolution of carbonate minerals. The third problem deals with the oxidation of pyrite in the vadose zone of mine tailings which result in acidic drainage conditions along with associated leaching of dissolved metals into the groundwater system. The results are presented to show the applicability and performance of the model to simulate the complex hydrogeochemical situations occurring in the three cases described above.

Key words : transport and geochemistry in groundwater, higher-order finite volume method, implicit and explicit schemes, acid mine drainage.

Introduction

The key chemical processes involved in subsurface transport are ion exchange, precipitation-dissolution, redox and acid base reactions. The simulation of these processes together in groundwater is practiced in recent years by combining the transport and geochemical models. Mangold and Tsang, (1991) present a review of various geochemical and transport models used in groundwater. Development of a hydrogeochemical model for solving multi-component solute transport with above specified chemical reactions depends on the choice of the transport model and the geochemical model and the nature of coupling between them. A large number of models have been developed for solving the advection and dispersion transport equation wherein the models using finite difference and finite element numerical methods are popular. In the last decade Eulerian-Lagrangian based models have been attempted to model advection and dispersion dominated situations. These models use separate numerical methods for the advective part and the dispersive part of the transport equation by using an operator split approach. In recent years there has been an interest to use higher order numerical methods such as finite volume and flux corrected transport approaches, developed in the computational fluid dynamics literature, for capturing the sharp fronts arising in the solution of the transport equation.

The popular geochemical models used in the literature are WATEQ, PHREEQE and MINTEQ. They differ among themselves in allowing the number of minerals and chemical constituents and type of chemical reaction models. MINTEQ uses Newton-Raphson method to get simultaneous solution of the nonlinear mass action expressions and linear mass balance relationships. The transport and geochemical models are coupled and solved sequentially using either an iterative or no iteration approach. Iterative approach is more popular but computationally intensive.

The present paper presents a numerical model which uses a finite volume approach for the transport equation coupled with the MINTEQ geochemical module. Finite volume methods differ among themselves in the choice of time stepping procedure used viz. implicit or explicit. In addition, finite volume methods are originally developed for solving the advective transport equation. Hence there is a need for analyzing the suitability of these methods for solving the advection-dispersion transport equation with and without operator-split framework. For this purpose, the numerical model is tested to study the performance on ion-exchange and precipitation-dissolution problems in dispersion dominated cases, while using (i) an explicit and implicit time stepping procedure, (ii) with and without operator split framework, and (iii) sequential iterative procedure (SIA) vis-à-vis non iterative procedure. The numerical model is applied for analyzing the leachate concentrations into groundwater during the oxidation of pyrite.

Governing Equations

The governing equations for the multicomponent transport in a multidimensional groundwater flow system with ion-exchange, sorption, precipitation and redox reactions, considering total analytical concentration (C_{Tj}) as the dependent variable, can be expressed as

$$\theta \frac{\partial C_{Tj}}{\partial t} + V.\nabla C_j = \nabla.\theta D.\nabla C_j$$

(1)

in which

$$C_j = c_j + \sum_{i=1}^{M_f} a_{ij} x_i \qquad j = 1, 2, \ldots, k$$

$$C_{Tj} = C_j + S_j + P_j$$

$$S_j = \sum_{i=1}^{M_y} a^y_{ij} y_i + \sum_{i=1}^{M_z} a^z_{ij} z_i$$

$$P_j = \sum_{i=1}^{M_p} a^p_{ij} p_i$$

The equation for species undergoing ion-exchange is

$$K_{ij} = \left(\frac{S_i}{C_i}\right)^{z_i} \left(\frac{C_j}{S_j}\right)^{z_j}$$

where K_{ij} is the selectivity coefficient. The equation for formation of precipitate P_{ij} is

$$C_i + C_j \rightarrow P_{ij}$$

solubility   product

$$Ks_{ij} = \frac{1}{[C_i][C_j]}$$

The equation for species undergoing oxidation and redox reactions is

$$bB_{red} + cC_{ox} \rightarrow dD_{ox} + gG_{red}$$

half reactions :

$$bB_{red} \rightarrow dD_{ox} + ne^-$$

$$gG_{red} \rightarrow cC_{ox} + ne^-$$

In the above equations t is time (T), C_{Tj} is total analytical (dissolved, sorbed and precipitated) concentration (M/L^3), S_j is total sorbed (adsorbed and ion-exchanged) concentration (M/L^3), P_j is total precipitated concentration (M/L^3), C_j is total dissolved concentration (M/L^3) of the jth aqueous component, V is Darcy velocity (L/T), k is number of aqueous components, c_j is the concentration of jth aqueous component species (M/L^3), x_i is the concentration of ith complexed species(M/L^3); a_{ij}^x is the stoichiometric coefficient of jth aqueous component in the ith complexed species. M_x is the number of complexed species; y_i is the concentration of the ith adsorbed species (M/L^3); a_{ij}^y is the stoichiometric coefficient of the jth aqueous component in the ith adsorbed species; M_y is the number of adsorbed species; z_i is the concentration of the ith

ion-exchanged species (M/L^3); a_{ij}^z is the stoichiometric coefficient of the jth aqueous component in the jth ion exchanged species; M_z is the number of ion-exchanged species; p_i is the concentration of the ith precipitated species (M/L^3); a_{ij}^p is the stoichiometric coefficient of the jth aqueous component in the ith precipitated species; M_p is the number of precipitated species.

Solution Procedure

The system of equations (1) are expressed using an operator-split approach for advection and dispersion parts as follows :

Advective transport

$$\frac{\partial C_{Tj}}{\partial t} = L_1(C_{Tj}) - L_1(S_j + P_j) \tag{5}$$

Dispersive transport

$$\frac{\partial C_{Tj}}{\partial t} = L_2(C_{Tj}) - L_2(S_j + P_j) \tag{6}$$

where

$$L_1 = -v_i \frac{\partial}{\partial x_i} \qquad L_2 = \frac{\partial}{\partial x_i}(D \frac{\partial}{\partial x_i}) \tag{7}$$

For operator-split approach Eqs.(5) and (6) are solved while using no operator-split approach Eq.(5) alone is solved by substituting (L_1+L_2) in place of operator L_1. The algorithm for the two step iterative approach, marching from time step t to a new time step (t+Δt), is

a) Solve Eq.(5) using explicit/implicit finite volume method and obtain $C_{Tj}^{\kappa+1}$, using C_{Tj}^t, S_j^t, P_j^t, C_{Tj}^κ, S_j^κ, P_j^κ. (Here κ referes to iteration number. In the first iteration, concentrations at t are used as previous iteration values).

b) The values of $C_j^{\kappa+1}$, $S_j^{\kappa+1}$, $P_j^{\kappa+1}$ are updated using MINTEQA2 based on $C_{Tj}^{\kappa+1}$ computed in step (a).

c) Check the convergence of $C_{Tj}^{\kappa+1}$ between successive iterations κ and

$\kappa+1$ for each node. If the convergence is not achieved repeat the above steps.

Model Testing

The finite volume model coupled with MINTEQ is tested using following cases:

Case 1: implicit time stepping with operator-split and SIA approach.
Case 2: implicit time stepping with no operator-split and SIA approach.
Case 3: explicit time stepping with no operator-split and SIA approach.
Case 4: no iterative approach with low Courant number (u Δt/Δx) of 0.2.
Case 5: no iterative approach with high courant number of 0.8.

Two problems are analysed using the above cases and model validation is done with *exact* solutions. As there are no exact analytical solutions for these test problems, solutions obtained from fine grid using a central difference method which is second order accurate in space and time are termed as the *exact* solutions.

Problem 1: Ternary ion-exchange

In this problem one dimensional transport of three ions ammonia, sodium, and calcium undergoing competitive ion exchange is presented. Such a scenario results during the remediation of mined area enriched with ammonium carbonate-bicarbonate in the ground water using successive sweeps of NaCl and CaCl$_2$ (Charbeneau, 1982).

Initially the leach solution in the mine area contains concentrations of ion1 (NH_4^+) and ion2 (Na^+) each equal to 100 meq/l. The chemical sweep to flush this solution consists of injection of one pore volume of NaCl solution at 200 meq/l followed by NaCl and CaCl$_2$ each at 100 meq/l. The domain is divided into 25 cells and grid size and time step are respectively 1m and 0.25 day with Courant number 0.2. The selectivity coefficients are $K_{12} = 4.5$ and $K_{13} = 9e-3$ l/meq.

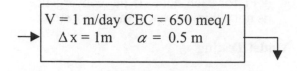

$$V = 1 \text{ m/day} \quad CEC = 650 \text{ meq/l}$$
$$\Delta x = 1m \quad \alpha = 0.5 \text{ m}$$

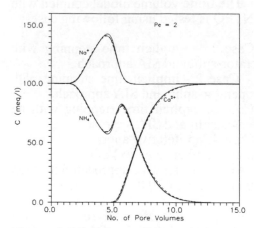

Figure 1: Comparison of numerical and exact solutions for ternary ion exchange. Exact solution: continuous line, case 1 (----) and case 4 (-.-.-.).

This problem is simulated using cases (1) and (4). The results for these cases are presented in Fig.(1) and compared with exact solution. It is noted from the figure that both cases could simulate the chromatographic behavior observed during the ion exchange process based on the affinity of Ca^{2+}, NH_4^+ and Na^+ with soil. The choice of using implicit time stepping with iterative approach produces accurate results. However, if no iteration is used between transport and geochemical models, the fronts are in agreement with the iterative approach at smaller time steps. Simulating the same case with case (5) produced results showing some smearing in the solutions of the fronts indicating the effect of courant number on the solution. It is observed that while using case(1) for a higher courant number did not make any deviations with exact solution suggesting that a choice of time step subjected to a Courant number less than 1 is not sensitive while using iterative approaches.

Problem 2: Simulation of multiple precipitation fronts:

In this problem complex changes resulting from fluid flow, multiple precipitation-dissolution fronts formed by precipitation and dissolution of minerals are presented. The problem is discussed by Yeh and Tripathi, (1991).

Table 1:

Chemical Species :
$CaCO_3$, $CaHCO_3^+$, $CaSO_4$, $CaOH^+$, $MgCO_3$, $MgHCO_3^+$, $MgSO_4$, $MgOH^+$, HCO_3^-, H_2CO_3, HSO_4
Minerals :
Calcite, $MgCO_3$, Gypsum, Calcium Hydroxide, $Mg_5(OH)_2(CO_3)_4$, $Mg(OH)_2$, $MgSO_4$

A one dimensional column is discretized into 100 cells. The porosity of the medium is 0.3 and liquid enters the column with a Darcy velocity of 0.5 dm/day. The simulation period is 250 days. Five chemicals are considered, calcium, magnesium, carbonate, sulfate and hydronium. The 12 aqueous species and minerals considered are shown in Table(1).

The initial concentrations of the chemicals are defined in such a way as to allow precipitation of magnesium carbonate mineral. The pH is fixed throughout the column between 7.7 and 8. The initial calcium concentration is constant at 10^{-4} M. The initial magnesium concentration decreases from 5.0×10^{-3} M at the left end of the column to 10^{-3} M at the right end of the column. The initial sulfate concentration decreases from 2×10^{-3} M to 10^{-3} M from left to right. But the initial carbonate concentration increased from left to right.

The boundary conditions for the chemicals entering left end of the column are 2×10^{-3}M, 10^{-3} M, 2×10^{-3} M for carbonate, magnesium and sulfate respectively. A calcium pulse of concentration 9×10^{-3} M is sent for first 9.5 days and is maintained constant at 1×10^{-4} M thereafter. This calcium pulse is introduced so as to induce supersaturation of carbonate minerals with space and time as the pulse travels.

Initially magnesium carbonate precipitation occurred in most of the column. The calcium pulse develops calcium carbonate precipitation at left end. The width of the calcium carbonate precipitation front increases and then decreases within the column due to the effect of increased input of the calcium pulse for finite duration from the input boundary. The dissolution of magnesium carbonate occurres at the left end of the column because of competition for carbonate from calcium.

Magnesium Carbonate precipitation at 190 days forms 5 precipitation fronts in the column. Hence at this time of 190 days the formation of multiple fronts obtained using different approaches are presented, compared and discussed. In Fig.(3), cases (1) and (2) are compared with exact solution. In Fig.(4), cases (2) and (3) are compared with exact solution. In Fig.(5), cases (4) and (5) are compared with exact solution. Figure (3) compares the cases of implicit finite volume using with and without operator-split approaches. It is observed that both the cases simulate the formation of the precipitation front locations and are in reasonable agreement with the exact solution. It is observed that the operator-split approach of case (1) captures the peaks of the precipitation fronts better than while using no operator-split approach within the implicit finite volume framework. This may be expected as the finite volume is more well suited for advection dominated cases and using this approach for simulating both the advective and dispersive fluxes together in a dispersion dominated situation may have relatively less accurate results. This result is interesting considering that the studies in the literature indicate that operator-split approach is less accurate while using an explicit finite volume method when tested on advection dominated settings.

Figure (4) compares the cases (2) and (3) pertaining to the implicit and explicit finite volume methods respectively while adopting no operator-split approach. It is observed that explicit finite volume produces inaccuracy in capturing the front location as well as the peak values of the fronts. It is interesting to note that the explicit and implicit finite volumes produce similar results for the ion exchange problem 1 whereas deviations between them is noted while simulating the more complex precipitation problem 2.

Figure (5) compares the cases (4) and (5) wherein the transport model and the geochemical model are not iterated within each time step. It is observed that no iterative approach produces numerical smearing which results in merging of the first and second precipitation fronts as they are close to each other. This behavior is same even when using a lower Courant number. However, case (4) which uses lower Courant number captures the peaks of the precipitation fronts. It is observed that the case (4) gives good results for the ion exchange problem while inferior results in the complex precipitation problem 2. Hence it is important to use an iterative approach when complex chemical reactions are involved. This aspect is to be noted as there are models developed recent times such as RAFT (Chilakapati et al., 2000) which uses no iterative approach.

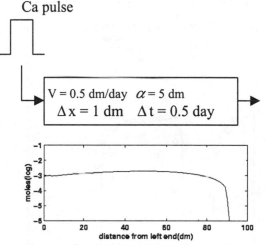

Figure 2: Initial Magnesium Carbonate precipitation in the column.

Model Application :

The model developed above is applied to analyse the leachate concentrations to the groundwater arising due to oxidation of pyrite. The data collection and analysis of this acid mine drainage problem has been studied in detail and reported in the literature

129

(Wunderly et al., 1996; Gerke et al., 1998). The problem addresses a situation wherein the top portion of the deposit of pyrite tailings get oxidized due to development

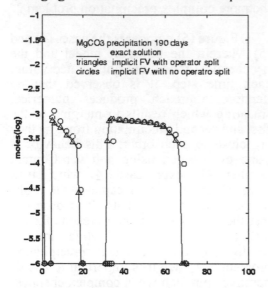

Figure 3: Distribution of Magnesium Carbonate precipitation in the column using SIA approach. case (1): Δ and case (2): o.

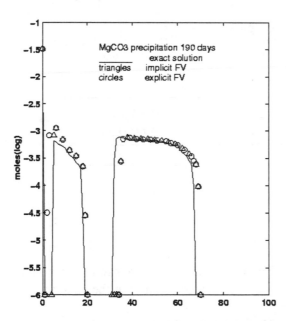

Figure 4: Distribution of Magnesium Carbonate precipitation using SIA approach and no operator split algorithm. case (1): Δ and case (3): o.

of unsaturated conditions in this zone. Initially the whole pyrite deposit is in saturated condition, which inhibits

oxidation of pyrite through abiotic or biotic processes due to low dissolved oxygen. The oxidation of pyrite particles results in the release of H^+, SO_4^{2-}, Fe(II) and Fe(III) and vertical transport of these components through the unsaturated and saturated layers of pyrite deposit and finally leaching into the groundwater. During the vertical transport of these components through the pyrite deposit, geochemical reactions take place resulting in release of additional chemical components and the transformation of all the components which depends on the mineral settings of the deposit. This phase results in the alteration of exisiting minerals as well as formation of new minerals which are sensitive to the concentrations of the transporting components. The oxidation of pyrite will eventually cease when all the pyrite content in the unsaturated zone gets exhausted and at that time the leachate concentrations reach equilibrium conditions. It is of interest to analyse the variation of concentration with depth of these components during their transport and transformation, when the pyrite is undergoing oxidation. In addition, it is of interest to analyse the temporal variation

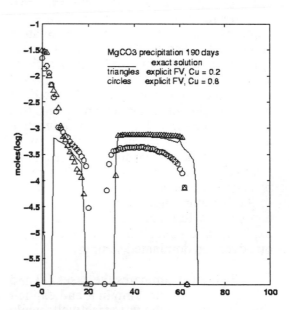

Figure 5: Distribution of Magnesium carbonate precipitation using no iterative approach and operator split algorithm. Case (1): continuous line, case (2): Δ and case (3): o

of the leachate concentrations to the groundwater until steady-state conditions are reached. One of the key chemical component for these analyses is H^+ concentration which reflects the pH content of the leachate.

A constant recharge influx of 0.31 m/yr is considered at the top of the oxidizing layer of pyrite for the total period of simulation. The porosity of the soil and the moisture content of the unsaturated zone are assumed time invariant as well. A porosity of 0.5 and a longitudinal dispersivity of 0.1m are considered. The solid phase components and geochemical reactions considered are shown in Table 2.

Table 2

Chemical components:
Ca^{2+}, Mg^{2+}, K^+, Cl^-, Al^{3+}, SO_4^{2-}, Fe^{3+}, CO_3^{2-}, H^+, Fe^{2+}, H_4SiO_4

Minerals :
Calcite, Siderite, Gibbsite, Ferrihydrite, Jarosite, Lime, Gypsum

$Mg^{2+} + CO_3^{2-} \Leftrightarrow MgCO_3^0$
$Mg^{2+} + CO_3^{2-} + H^+ \Leftrightarrow MgHCO_3^+$
$Mg^{2+} + SO_4^{2-} \Leftrightarrow MgSO_4^0$
$Ca^{2+} + CO_3^{2-} \Leftrightarrow CaCO_3^0$
$Ca^{2+} + SO_4^{2-} \Leftrightarrow CaSO_4^0$
$K^+ + SO_4^{2-} \Leftrightarrow KSO_4^-$
$Al^{3+} + 2H_2O \Leftrightarrow Al(OH)^{2+} + 2H^+$
$Fe^{2+} + H_2O \Leftrightarrow FeOH^+ + H^+$
$Fe^{2+} + SO_4^{2-} \Leftrightarrow FeSO_4^0$
$CO_3^{2-} + H^+ \Leftrightarrow HCO_3^0$
$CO_3^{2-} + 2H^+ \Leftrightarrow H_2CO_3^0$
$Al^{3+} + H_2O \Leftrightarrow AlOH^{2+} + H^+$
$Al^{3+} + SO_4^{2-} \Leftrightarrow AlSO_4^+$
$Al^{3+} + 2SO_4^{2-} \Leftrightarrow Al(SO_4)^{2-}$
$H^+ + SO_4^{2-} \Leftrightarrow HSO_4^-$
$Fe^{3+} + H_2O \Leftrightarrow FeOH^{2+} + H^+$
$Fe^{3+} + SO_4^{2-} \Leftrightarrow FeSO_4^+$
$Fe^{3+} + 2H_2O \Leftrightarrow Fe(OH)^{2+} + 2H^+$
$Fe^{3+} + 2SO_4^{2-} \Leftrightarrow Fe(SO_4)^{2-}$
$Fe^{3+} + e^- \Leftrightarrow Fe^{2+}$
$CO_3^{2-} + 2H^+ \Leftrightarrow CO_2(g) + H_2O$

The reaction equations during the oxidation of pyrite are as follows:

$FeS_2 + H_2O + 7/2O_2 => Fe^{2+} + 2SO_4^{2-} + 2H^+$
$FeS_2 + 1/2H_2O + 15/4O_2 => Fe^{3+} + 2SO_4^{2-} + H^+$

Simulation results

The implicit finite volume method with SIA approach is used to perform simulations for a period of 12 years for this field problem. The results of the simulations are compared with the field measured concentrations reported by Wunderly et al., (1996) for analyzing the performance of the numerical model.

Figure (6) presents the measured and simulated concentrations of pH, Cl^-, Mg^{2+}, Al^{3+}, SO_4^{2-} and Fe(total) at various depths of the pyrite deposit at the end of 12 years. It is observed from the figure that the numerical model simulates the general trend of concentration fronts. In the vadose zone, the pH is buffered by jarosite precipitation and dissolution at a pH of about 2.0. Iron concentration shows two distinct peaks (Figure 6). The upper peak in the ferrihydrite buffered zone consists mostly of Fe(III) and in the lower peak where gibbsite is the buffering mineral consists mostly of Fe(II). Below depth of 3-m the modeled and field results of Fe(total) concentrations differ substantially. Possible explanations for the discrepancy may be the choice of a simplified numerical model wherein abiotic processes and equilibrium chemistry alone are considered.

Sensitivity analysis

The numerical model is used to study the sensitivity of the Henry's constant during the pyrite oxidation as this is a critical parameter which effects the diffusion of oxygen from air to pore water. This Henry's constant varies widely for various temperatures. To analyse this effect, Henry's constant pertaining to cold and relatively warmer geographical regions are considered. A mean annual temperatures of $6^\circ C$ and $20^\circ C$ are considered for cold and warmer regions respectively. It is to be noted that the simulation results shown in Fig.(6) are for a temperature of $10^\circ C$. Figure (7) presents the concentrations of pH and Fe (total) at

131

various depths for these two Henry's constants. It is observed from the figure that, using Henry's constant associated with higher temperatures result in higher pH concentrations as expected at the oxidizing zone due to lower dissolved oxygen content. The variation in pH along the depth for these cases increases. It is also observed that variation of pH concentration with depth is negligible at larger depths for both the cases. This implies that the pH of the leachate during acid mine drainage situations is higher in warmer regions in comparison with colder regions. The variations of Fe(total) concentrations with depth correlate with the variations of pH due to changes in Henry's constant indicating higher Fe(total) concentrations associated with higher oxidation states and associated low pH values for a lower Henry's constant. However, it is interesting to note the Fe(total) front mobility and spread is nearly same even though Henry's constants are varied in this complex reaction setting. Figure (8) presents the temporal variations of pH, Cl$^-$ and Fe^{2+} concentrations entering the groundwater at the bottom of the pyrite oxidation zone (1.1m from top) using higher and lower Henry's constants. It is observed from the Fig.(8) that, Fe^{2+} increases after the start of pyrite oxidation and reaches a peak value after certain period of time and later decreases due to reduction in the production of sulphate and iron from the pyrite oxidizing zone. The pH concentration starts decreasing from the higher initial values due to the production of acid water from the oxidation zone. It is also noted that the pH decrease follows a step like drops indicating that between two steps the decrease is monotonic. The constant leaching concentration of pH for certain time may be attributed to the buffering reactions during that time period which maintain a constant pH. After a long time the leaching concentrations tend to reach equilibrium with time which can be noted from the Fe^{2+} behavior. The temporal variation in Cl$^-$ concentrations results from the difference in the inflow and initial chloride concentrations. The results of Cl$^-$ are presented to give an idea of the temporal front movement of a conservative component. It is also

observed from the Fig.(8), that the effect of increased Henry's constant is to show smaller peak in Fe and a higher pH in the leachate. The decrease of pH with steps is also noted for higher and lower Henry's constants. However, it is observed that the Fe front experiences a lag for a higher Henry's constant implying a longer time needed for Fe concentrations to reach steady state.

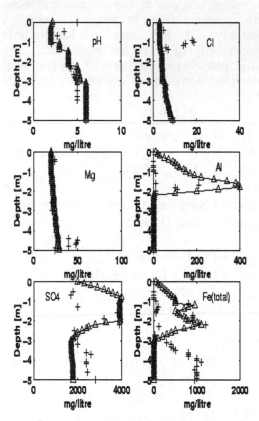

Figure 6: Simulation results, Field data: +, literature: ___, case 1: Δ

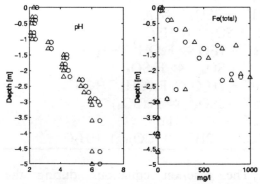

Figure 7: Distribution of pH and Fe(total) for low (triangles) and high (circles) Henry constants

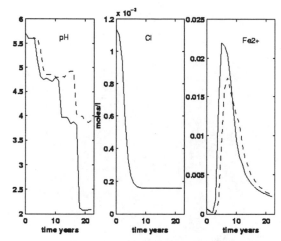

Figure 8: Fluxes entering saturated zone for low (continuous line) and high (dashed line) Henry constant for 22 years simulation.

Conclusions

A numerical model based on finite volume method is presented for solving transport and geochemistry in groundwater. The model is tested for implicit and explicit time stepping procedures, effect of using operator split approach and using no iterative approach between transport and geochemical modules. For this purpose two test problems involving ion exchange and precipitation/ dissolution are used. It is observed that implicit finite volume is better in capturing the fronts peclet which can only be observed in the precipitation problem. Use of operator split approach in less sensitive. Use of no iterative approach produces inaccurate results even at low Courant numbers for dispersion dominated settings and once again noted in the smearing produced in the precipitation problem. The model is applied for field problem of oxidation of pyrite and resulting leachate concentrations to the groundwater.

References

Appelo, C.A.J. and Postma, D., 1994. Geochemistry : Groundwater and Pollution, A.A.Balkema, Netherlands. 535pp

Ashok Chilakapati, Yabusaki, S., Szecsody, J., MacEvoy, W., 1999. Groundwater flow, multicomponent transport and biogeochemistry: development and application of a coupled process model, J. Contaminant Hydrology, 43:303-325.

Charbeneau, R.J., 1981. Groundwater Calculation of Pollutant Removal During Groundwater Restoration with Adsorption and Ion Exchange. Water Resour. Res., 1117-1125.

Gerke, H.H., Molson, J.W., Frind, E.O., 1998. Modelling the effect of chemical heterogeneity on acidification and solute leaching in overburden mine spoils. J. Hydrology, 209:166-185.

Mangold, D.C., and Tsang, C.F., 1991. A summary of subsurface Hydrological and ydrochemical models. Rev. Geophys., 29(1): 51-79

Walter, A.L., Frind, E.O., Blowes,D.W., Ptacek, C.J., and Molson, J.W., 1994. Modeling of multicomponent reactive transport in groundwater. 1. Model development and evaluation. Water Resour. Res., 30(11): 3137-3148.

Wunderly, M.D., Blowes, D.W., Frind, E.O., and Ptacek, C.J., 1996. Sulfide mineral oxidation and subsequent reactive transport of oxidation products in mine tailings impoundments: A numerical model, Water Resour. Res. 32(10): 3173-3187.

Yeh, G.T., and Tripathi, V.S., 1989. A critical evaluation of recent developments in hydrogeochemical transport models of reactive multichemical components. Water Resou. Res., 25(1): 93-108.

Yeh, G.T., and Tripathi, V.S., 1991. A model for simulating transport of reactive multispecies components : Model development and demonstration. Water Resour. Res., 27(12): 3075-3094.

Intl. Conf. on Sustainable Development and Management of Groundwater Resources
in Semi-Arid Region with Special Reference to Hard Rock, (IGC 2002),
M. Thangarajan, S.N. Rai & V.S. Singh (Eds.)

Evolving pre-development management schemes through mathematical modelling: A case study in Kunyere river valley, Okavango delta, Botswana

M. Thangarajan

National Geophysical Research Institute, Hyderabad-500007, India, mthangarajan@satyam.net.in or
mthangarajan@hotmail.com

Abstract

The over development of groundwater resources leads to the decline of water level causing socio-economic and environmental problems. It is, thus, imperative to manage the groundwater resources in an optimal manner. Management schemes can be evolved, only if the groundwater potential is assessed in more realistic manner. Mathematical modelling in conjunction with detailed field investigations have been proved to be a potential tool for this purpose. Evolving pre-development management schemes is still works out to be better choice. One such study was carried out in Kunyere River valley, Okavango Delta, Botswana. Kunyere River valley has three tributaries viz. Marophe, Xudum and Matsibe Rivers. The valley falls in the southeastern fringe of Kalahari Desert (Botswana), which tends along the Kunyere fault in a northeast to southwest direction. The materials in the valley system are saturated below a depth of 7 to 9 m below ground level (bgl), where a fresh potable groundwater reserve is present to a maximum depth of 50 to 70 m (bgl) and below this depth, groundwater is brackish one. The Department of Water Affairs, Govt. of Botswana, has quantified the groundwater resource in the valley through exploratory drilling, test pumping, and hydro-chemical analysis of groundwater samples.

A model with six layers of flow regime was conceptualised by making use of available data. Fourth layer of the aquifer system is the main fresh water bearing aquifer and the bottom layer is brackish one. Long duration pumping test carried out in this area indicated the leaky nature of the aquifer system. Mathematical model of the basin was constructed and calibrated for steady state condition by using Visual Modflow computer software. Two prognostic runs were made and an optimal one was identified which will ensure minimum upward leakage from the bottom saline unit to the pumping aquifer. This simulation study indicates that substantial development of groundwater potential is possible in this area.

Key Words: Kunyere River valley, Okavango Delta, Mathematical Modelling, Management schemes, Botswana

Introduction

Okavango Delta (Fig. 1) is located in Botswana (Southern Africa). The Okavango River, which originates in Angolan highlands reaches in northwestern Botswana and terminates as a huge inland delta, namely the Okavango Delta. The Kunyere River valley runs along the Kunyere Fault and drains the basin to Lake Ngami. Maun a small Town, which is located on the bank of river Thamalaknae is emerging as a famous wild life tourist centre in Botswana and the demand for more drinking water, is ever increasing. The demand during the year 1997 was about 3540m^3/day and it is likely to increase to about 4200m^3/day during the year 2000. The present demand is met from the well fields located in the valleys of Shashe and Thamalakane and the future demand has to be met from new resources. Therefore, Government of Botswana formulated a Project in the year 1996 to explore and assess the new groundwater resources through integrated geophysical and geohydrological studies in the neighbourhood of Maun. The Kunyere River valley covering an area about 120 km^2 is one such system wherein geohydrological, geophysical and chemical quality studies were combined to find suitable sources for exploiting groundwater. The data collected under this

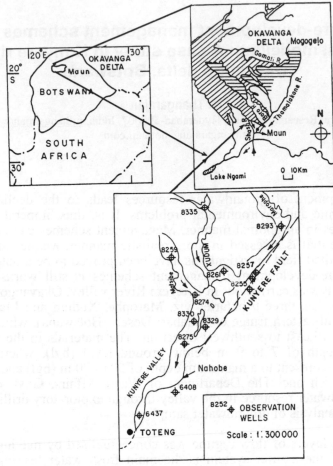

Figure 1: Location Map of Kunyere River Valley

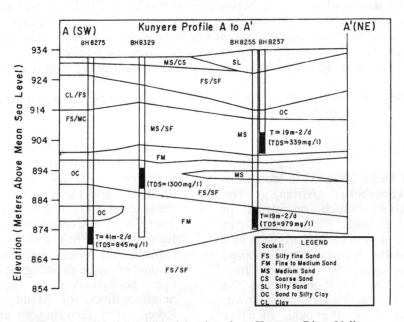

Figure 2: Vertical Geological Section along Kunyere River Valley

136

project was used to conceptualise the groundwater flow regime and a preliminary mathematical model was developed to study the aquifer response and thereby to evolve pre-developmental management schemes. The model was constructed and calibrated for steady state condition. The model calibration has clearly indicated that upper and middle aquifers are semi-confined in nature. The calibrated model was then used to study the aquifer response under two possible cases for evolving optimal well field locations in the upper semi-confined aquifer.

Hydrogeological setting

The aquifer system belongs to Kalahari beds. Three exploration boreholes and one water-level observation borehole have been drilled within the Kunyere valley. The vertical subsurface geological section along the valley is shown in Fig. 2. The lithology encountered during the drilling of these boreholes indicates a thin surficial cover of silts ranging in thickness from less than one meter in the south west (BH8275) to about seven metres in the northeast at the location of BH8255/BH8257 (Fig. 2). Clayey fine sands and silty fine sands underlie these silts with a thickness ranging from 15 to 20 m. The unit is in turn underlain by medium sand and fine to medium sand with a thickness between 11 to 19 meters. The medium sand and fine to medium sand is underlain by sandy to silty clay and silty fine sand with a thickness varying between 8 to 15 m. This unit is underlain by fine to medium sand interbedded with clay lenses (BH8275), which has a thickness varying from 7 to 21 meters. The fine to medium sand increases in thickness from the northeast to southwest down the Kunyere Valley and is underlain by fine silty sand. There are two main aquifer systems in the river valley (Fig. 2). The top zone (surface soil) is the low permeable zone with average thickness of 10 m. The top two aquifers are fresh water bearing units and the bottom one is saline. An outline of the hydrogeological conditions in each of these units is given below.

Upper Semi-confined Aquifer (Layer 2)

The upper semi-confined aquifer consists of fine to medium grained sand. The upper aquifer has a more or less uniform thickness throughout the reach of the valley between BH8255 and BH8275. This upper aquifer unit occurs at approximately 20 m below ground surface (approximately 910 m (amsl)) and the bottom occurs at around 36 m below ground surface (approximately 895 m (amsl)). The thickness of this aquifer is about 16 m. Recharge to the semi-confined aquifer has been predominantly from river water infiltration as downward leakage. Major rainfall events also recharge the top aquifer to a limited extent and it is derived from model calibration as 1.8 mm per annum. This is well within the range of value recommended by the recharge committee of Botswana. The mean annual rainfall for 1986-1996 from the Maun airport weather station is about 380 mm. Presently the main discharge from this aquifer is due to evapotranspiration.

Lower Semi-confined Aquifer (Layer 4)

The lower aquifer occurs between 45 and 68 m below ground surface (approximately 890 m (amsl) to 880 m (amsl)) and appears to thicken in a south-western direction. In general this aquifer has lenses of finer grained material (silts and silty clays) in the south near the Matsibe junction. The average thickness of this aquifer is about 24 m. Lateral extent or boundaries of the fresh water aquifer in the Kunyere Valley have been delineated on the basis of an air-borne Electro-magnetic survey flown over the area as well as transient Electro-magnetic (TEM) soundings. This survey indicates that the lateral extent of the fresh water within the Kunyere valley varies from 5.5 km in the southwest near the Kunyere/Matsibe junction to approximately 3 km in the northeast near the Kunyere/Marophe junction. This reach of the Kunyere valley covers a total length of approximately 30 km. The average area of the fresh water aquifer covered by this stretch of the Kunyere Valley is approximately 127 km^2. The depth of fresh water in the Kunyere

Valley was delineated on the basis of TEM soundings conducted in the valley. A contoured depth to interpreted fresh water/saline water interface based on TEM soundings indicates that fresh water occurs to a maximum depth of 70 m and minimum depth of approximately 50 m.

The transmissivity (T) values were estimated through single well pumping tests in the upper and lower semi-confined aquifers. T value of 19 m^2/day was estimated in layer 2 and 41 m^2/day in layer 4. The storativity values were inferred and not estimated. The initial water level contour is shown in Fig. 3.

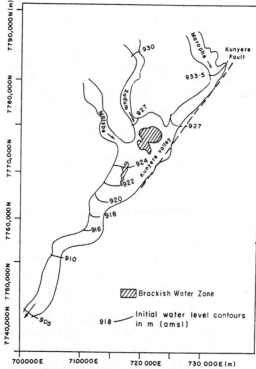

Figure 3: Initial water level contour in m (amsl) for lower semi confined aquifer

Bottom Aquifer (Brackish, Layer 6)

The aquifer, which occurs approximately between 70 and 75 m below the ground level, comprised of fine to medium sands and contains brackish to saline water. There is not much information available regarding hydraulic characteristics of this unit. The hydraulic conductivity value is assumed of 7 m/d and its storativity value as 2.9 x 10-4. The bottom aquifer is hydraulically connected

with the overlying middle semi-confined aquifer.

Model design

Methodology

Groundwater models are simplified representations of the subsurface aquifer systems. The calibrated and validated models may be used to predict aquifer response to hydraulic stresses three dimensional groundwater flow in a porous media such as pumping recharging et c. can be expressed by the following partial differential equation (Rushton and Redshaw, 1979):

$$\frac{\partial}{\partial x}\left(K_{xx}\frac{\partial h}{\partial x}\right)+\frac{\partial}{\partial y}\left(K_{yy}\frac{\partial h}{\partial y}\right)+\frac{\partial}{\partial z}\left(K_{zz}\frac{\partial h}{\partial z}\right)=S_s\frac{\partial h}{\partial t}\pm W$$

(1)

Where,

Kxx, Kyy, Kzz are the hydraulic conductivities along x, y, and z co-ordinates which are assumed to be parallel to the major axes of hydraulic conductivity (LT^{-1}), h is potentiometric head (L), W is volumetric flux per unit volume and represents sources and/or sinks of water (T^{-1}), S is the specific storage of the porous material (L^{-1}), and t is observation time (T)

Equation (1) describes groundwater flow under non-equilibrium conditions in a heterogeneous and anisotropic medium. The groundwater flow equation together with specification initial and boundary conditions constitutes a mathematical representation of the groundwater flow in aquifer system. In general numerical methods are used in general to solve this groundwater flow equation.

The computer software "Modflow", developed by the United States Geological Survey (USGS, 1988), was used for the present study. In this program is solved using the Block Centred Finite Difference Approach groundwater flow equation (1). A pre-and post-model processor viz. Visual-Modflow developed by Nilson Guigner and Thomas Franz of Waterloo Hydrologic Software Inc., Waterloo, Ontario, Canada (1996) was used for graphical data input, and for analysis and presentation of the output data.

Model configuration

The Kunyere River valley aquifer system was conceptualised as a six-layer system with three aquifers separated by two confining layers (confining/semi-confining) as desired below:

Top soil with low permeable zone(layer 1)
Upper semi-confined aquifer(Layer 2)
Upper confining unit(Layer 3)
Lower semi-confined aquifer(Layer 4)
Lower semi-confining unit (Layer 5)
Bottom brackish/saline aquifer(Layer 6)

The study area was divided in to square grids and the map is shown in Fig. 4.

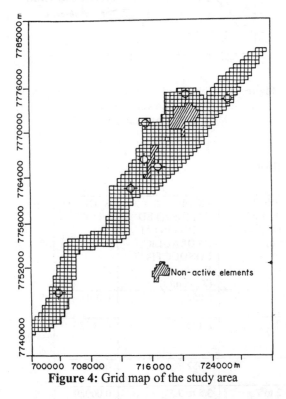

Figure 4: Grid map of the study area

The confluence points of three tributaries viz. Matsibe, Xudum and Morphe were taken as the inflow boundaries in the upper and lower semi-confined aquifers. The quantum of inflow flux was calculated by using transmissivity values and the hydraulic gradient. It was estimated that about 1100 m^3/d is received in the upper aquifer and lower aquifer. Layers 3 and 5 (silty sands and clays) were

taken as aquitards. Since bottom aquifer (saline unit) is laterally extended, the northeast and southwest boundaries were assumed as inflow and outflow boundaries respectively. The eastern and western lateral boundaries were treated as no flow boundaries as the flow is predominantly from north -east to south -west. The subsurface out flow towards the south-western direction was simulated as fixed heads near Toteng. The water levels monitored in the presently drilled bore wells are very useful in fixing the boundary heads in all the three layers. The aquifer parameters i.e. hydraulic conductivity, K (m/day), Specific yield (Sy) and Specific storage (Ss (L^{-1})) were assigned in zone wise for each layer. Storativity value of 0.00029 was uniformly assumed for the upper and lower semi-confined aquifers. The vertical permeability for each layer is assigned as one tenth of hydraulic horizontal conductivity. The upper semi-confined aquifer unit and lower confined aquifer unit were assumed to have storativity value of 0.001. The hydraulic conductivity value of 0.052 m/day was set for both the upper confined and lower confined aquifer.

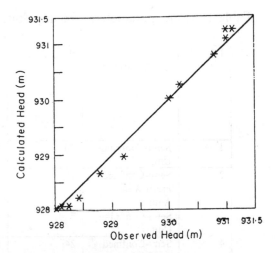

Figure 5: Comparison of Computed Vs Observed Heads (Steady State Model)

Model Calibration

Steady State Calibration

The aquifer condition of March 1997 was assumed to be the initial condition for the steady state model calibration. The model could not be initialised to an early date due to the non-availability of water level data before January 1997. Minimising the difference between the computed and the field water level for each observation point started the steady state model calibration. Number of trial runs were made by varying the input / output stresses and the hydraulic conductivity values of the top and middle aquifers in order to keep the root mean square (RMS) error below 0.2 m and mean error below 0.1 m. The computed versus observed head for selected observation points are compared in Fig. 5. This figure indicates that there is a fairly good agreement between the calculated and observed water levels. The computed water level contours in layer 4 and 6 are shown in Figs. 6 and 7 respectively. The calibrated zonal hydraulic conductivity (K) values for the upper, lower and bottom aquifers and upper aquitard are shown Table 1. Due to non- availability of historical water level data, transient state calibration could not be initiated in the present study. The calibrated steady state model was then used to progonose the aquifer response.

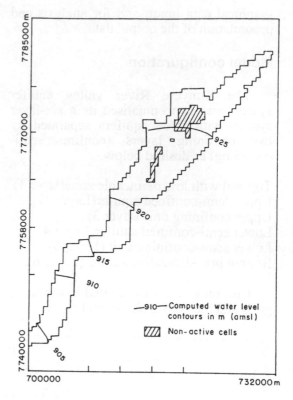

Figure 6: Computed Water Level Contours in layer 4 (Lower semi-confined aquifer) (Steady State, January 1997)

Table 1: Model parameters - Steady State condition

LAYER	UNIT DESCRIPTION	AVERAGE	FIELD DERIVED HYDRAULIC CONDUCTIVITY (K in m/d)	MODEL CALIBRATED HORIZONTAL HYDRAULIC CONDUCTIVITY (K_x in m/d)	STORATIVITY
1	Semi-Confining Unit	20	NA	0.052	0.001
2	Semi-Confined Fresh Water Aquifer	15	$K_x = 3.2$	2.1 to 3.2	0.00029
3	Semi-Confining Unit	7	NA	0.052	0.001
4	Semi-Confined Fresh Water Aquifer	24	$K_x = 3.2$ to 7	3.5 to 9.5	0.00029
5	Semi-Confining Unit	2	$K_x = 0.0055$	0.052	0.001
6	Semi-Confined Brackish Aquifer	50	NA	7	0.00029

K_x = Horizontal Hydraulic Conductivity. K_x = Vertical Hydraulic Conductivity, Assumed to be 0.1 of K_x.

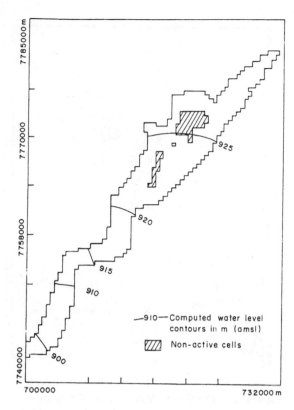

Figure 7: Computed Water Level Contours in layer 6 (Lower semi-confined aquifer) (Steady State, January 1997)

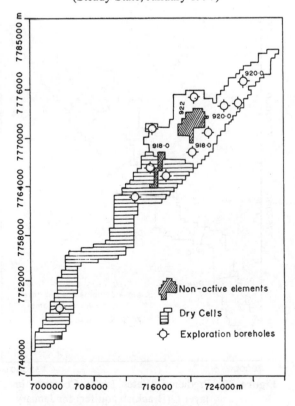

Figure 8: Location of the wells

Prediction scenario runs

Prediction of water level were made for two cases.

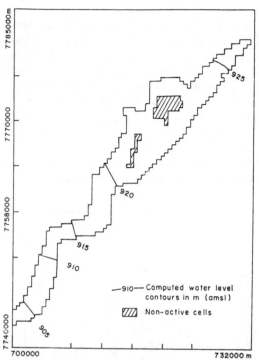

Figure 9: Predicted Water Level Contours in layer 4 (Lower Semi-confined aquifer) for January 1998 (Case-1)

Case 1

In case 1, 10 boreholes (Fig. 8) were placed with 3-km spacing over an area of 93 km^2 in the lower semi-confined aquifer, each pumping at a rate of 480 m^3/d over a 10-year period. Table 2 summarises the computed results for inflows and outflows to the lower semi-confined aquifer. The bulk of inflow to the lower semi-confined aquifer is through downward vertical leakage from the upper semi-confined aquifer and the upper semi-confined aquifer (Column 6 on Table 2). Contribution to pumping from aquifer storage in the lower semi-confined aquifer is insignificant (Column 8). There is a net vertical downward leakage to the lower brackish/saline aquifer from the lower fresh water semi-confined aquifer until year 8 in the simulation (Column 9). After 8th year, there is an 8 to 9 percent

141

contribution to pumping in the lower semi-confined aquifer due to upconing of water from the brackish/saline aquifer (Column 12). The impact of this upconing in 10th year of the simulation is an increase in TDS from a baseline of 800 mg/l to 1,147 mg/l.

Figures 9 and 10 show predicted water-level contours in lower semi-confined aquifer (Layer 4) and bottom brackish (Layer 6) aquifer for January 1998 (First year); Figures 11 and 12 show the same for January 2007 (Tenth year). These figures indicate that the dewatering of the top surface layer progresses northeastward over the simulation period. The water-level contours indicate an approximate 5 m decline over the 10-year simulation.

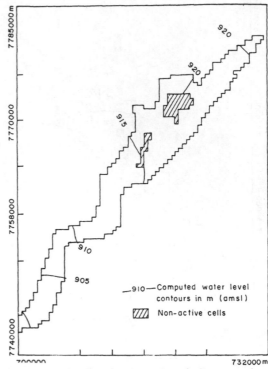

Figure 11: Predicted Water Level Contours in layer 4 (Lower Semi-confined aquifer) for January 2007 (Case-1)

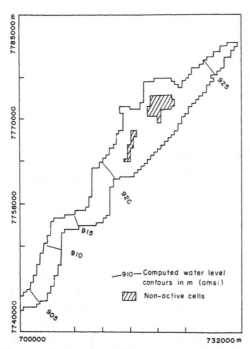

Figure 10: Predicted Water Level Contours in layer 6 (Brackish aquifer) for January 1998 (Case-1)

Figures 13 and 14 are computed water-level hydrographs for boreholes 8255, 8257 and 8274, 8275 respectively. Except one borehole 8255, all other boreholes are completed in the lower semi-confined aquifer (Layer 4) and the first one in the upper semi-confined aquifer (Layer 2). Water levels in the lower semi-confined aquifer appear to drawdown by 7 m at BH8255 (north-eastern portion of the

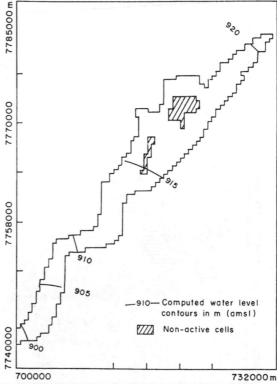

Figure 12: Predicted Water Level Contours in layer 6 (Brackish aquifer) for January 2007 (Case-1)

142

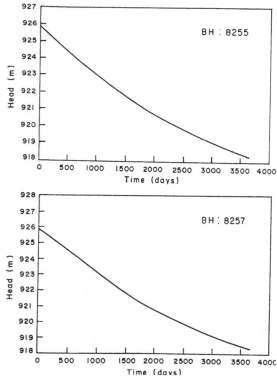

Figure 13: Predicted Well Hydrographs for Observation Wells 8255 and 8257 (Case-1)

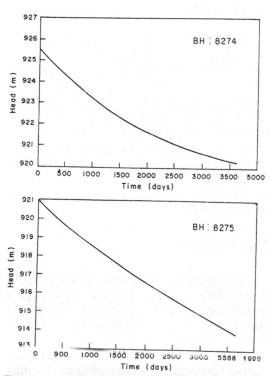

Figure 14: Predicted Well Hydrographs for Observation Wells 8274 and 8275 (Case-1)

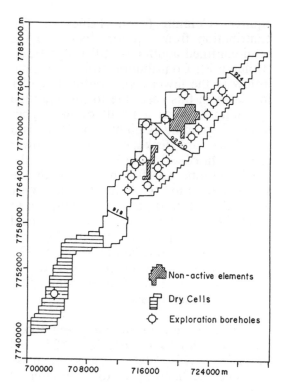

Figure 15: Location of the wells and predicted water level contours

valley) and at BH8275 (south-western portion of valley) after 10 years in this abstraction scenario. Also, there appear to be 7 m of drawdown within the upper semi-confined aquifer (Layer 2) at BH8257. The top layer (Layer 1) becomes dewatered from the Nhabe/Kunyere confluence to the lower end of the wellfield. The correlation of drawdown in the upper and lower aquifer systems implies significant leakage between these two units. Similar study has been carried out for case 2 with 20 number of boreholes as shown in fig. 15.

Case -2

In this case, the number of production boreholes placed in the lower semi-confined aquifer within the 93 km^2 area was doubled to 20 boreholes (1.5 km spacing), each pumping at 480 m^3/d over the 10 year simulated period. The inflows and outflows from the semi-confined aquifer for this scenario are summarised on Table 3. As in case 1, the bulk of the water supplied to pumping is via vertical downward leakage from the upper semi-confining layer and upper semi-confined

143

aquifer (Column 6 on Table 3). The contribution from aquifer storage in the semi-confined aquifer is still insignificant (Column 8). Contribution from horizontal inflow to the lower semi-confined aquifer increases in this case due to lowering the hydraulic head as a result of the higher rates of pumping (Column 7). There is a net downward leakage from the lower semi-confined aquifer to the brackish/saline aquifer in Year 8 of the simulation period (Column 9). After Year 8, there is a 5 to 6 percent contribution to pumping in the lower semi-confined aquifer due to upconing of water from the lower brackish/saline aquifer (Column 12). The calculated water quality impact due to this upconing in Year 10 of the simulation is similar to that in case 1.

Figures 16 and 17 are computed water-level hydrographs for BH8255, 8257 and BH8274, 8275. Water levels in these boreholes declined by 12 to 13 m after 10 years.

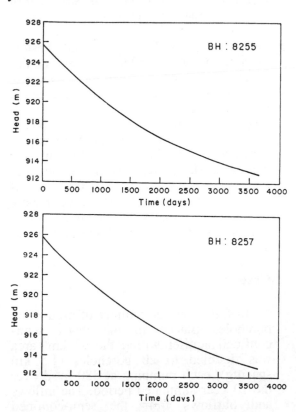

Figure 16: Predicted Well Hydrographs for Observation Wells 8255 and 8257 (Case-2)

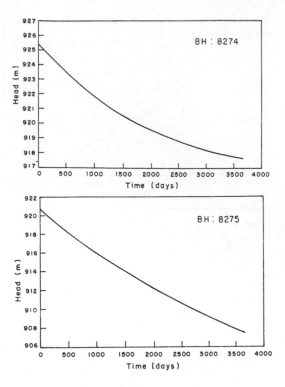

Figure 17: Predicted Well Hydrographs for Observation Wells 8274 and 8275 (Case-2)

Discussion

The preliminary modelling supports the future consideration of this area for development. It appears that the area can support a considerable amount of freshwater withdrawal without significant water quality changes as a result of upconing. As with most of the explored area, the hydraulic conductivity of the geologic materials that separate the fresh and brackish groundwater is an uncertainty of the model and needs to be investigated in more detail.

Conclusions

The present modelling study is very preliminary and subject to many assumptions and assumed hydraulic parameters. The infiltration due to river flow is not considered in this study, as there was no regular flow for the last three years. The preliminary modelling for this area indicates scope for potential development over a portion of this area. Simulations study indicates that the pumping of 4,800 and 9,600 m3/day from

144

10 and 20 borewells respectively, producing impacts to the top layer in terms of de-watering, were sustainable if this area receives regular river flow in the future.

Acknowledgements

The author wishes to thank the Government of Botswana for permitting to publish this paper. Authors also wish to thank Director, National Geophysical Research Institute (NGRI), Hyderabad, India for permitting the first author to visit Botswana as modelling expert to M/s. Eastend Investments (PTY.) Ltd., Gaborone, Botswana. Maun Groundwater Exploration Project Director Mr. David Ede and his team members are thanked for their assistance in providing the additional data required for this study. M/s M. Jayarama Rao, M. Kranti Kumar and P. T. Varghese at NGRI are thanked for their assistance in tracing and preparation of the manuscript.

References

Guiguer, N. and Franz, T., 1996. Visual Modflow, Waterloo Hydrogeologic Software, Waterloo, Ontario, Canada

Maun Water Supply Hydrogeological Survey, 1988. BRGM Technical Report, Vol. 1.

Michael G. Mcdonald and Arlen W. Harbaugh, 1988. A Modular three-dimensional Finite Difference Groundwater Flow Model, USGS Technical Report on MODELING Techniques, Book 6.

Rushton, K.R. and Redshaw, S.C., 1979. Seepage and Groundwater Flow, John Wiley and Sons Ltd., pp 1-330.

Shashe Well Field Conceptual Model, Interim Report (Appendix G), 1996 by Eastend Investments (PTY.) Ltd., Gaborone, Botswana.

Table 2: Model Prediction Scenario 1: Pumpage of 4800 m3/d from Lower Semi-Confined Aquifer with no River Recharge

Year	Pumping Rate from Lower Semi-Confined Aquifer (m³/d)	Horizontal Outflow (m³/d)	Downward Leakage between lower semi-confined aquifer and brackish/saline aquifer (m³/d)	Total Outflow from the lower Semi-Confined aquifer (m³/d)	Inflow from Downward Leakage from Upper Semi-Confining Layers and Upper Semi-Confined Aquifer (m³/d)	Horizontal inflow (m³/d)	Contribution from Aquifer Storage in the Semi-Confined Aquifer (m³/d)	Upward leakage Between lower Semi-Confined Aquifer and Brackish/Saline Aquifer (m³/d)	Total inflow to the Lower Semi-Confined Aquifer (m³/d)	Percent Contribution to Pumping from Lower Semi-Confined Aquifer Storage	Percent Contribution to Pumping from Upconing
1	4800	2288	3397	10,485	6179	1147	95	0	10,485	2	0
2	4800	2754	3578	11,132	6944	1310	108	0	11,132	2	0
4	4800	3169	1602	9571	6744	1535	94	0	9571	2	0
8	4800	4465	0	9265	6138	1852	68	394	9265	1	8
10	4800	4394	0	9194	5528	1974	50	431	9194	1	9

Table 3: model prediction scenario 2: Pumpage of 9600 m³/d from Lower Semi-Confined Aquifer with no river recharge

Year	Pumping Rate from Lower Semi-Confined Aquifer (m³/d)	Horizontal Outflow (m³/d)	Downward Leakage between lower semi-confined aquifer and brackish/saline aquifer (m³/d)	Total Outflow from the lower Semi-Confined aquifer (m³/d)	Inflow from Downward Leakage from Upper Semi-Confining Layers and Upper Semi-Confined Aquifer (m³/d)	Horizontal inflow (m³/d)	Contribution from Aquifer Storage in the Semi-Confined Aquifer (m³/d)	Upward leakage Between lower Semi-Confined Aquifer and Brackish/Saline Aquifer (m³/d)	Total inflow to the Lower Semi-Confined Aquifer (m³/d)	Percent Contribution to Pumping from Lower Semi-Confined Aquifer Storage	Percent Contribution to Pumping from Upconing
1	9600	2286	33125	15,011	9877	4948	186	0	15,011	1	0
2	9600	2735	3339	15,674	10,194	5295	185	0	15,674	1	0
4	9600	3153	1420	14,173	9342	4674	157	0	14,173	1	0
8	9600	4417	0	14,017	7925	5454	105	533	14,017	1	5.5
10	9600	4326	0	13,926	7060	6214	79	573	13,926	1	6

Intl. Conf. on Sustainable Development and Management of Groundwater Resources
in Semi-Arid Region with Special Reference to Hard Rock, (IGC 2002),
M. Thangarajan, S.N. Rai & V.S. Singh (Eds.)

Modeling of groundwater flow in Schoneiche waste disposal site, Germany

S.N. Rai

National Geophysical Research Institute, Hyderabad 500 007, India
(Email: postmast@csngri.ren.nic.in)

Abstract

In order to find out some ways and means to prevent or at least minimize the spreading of hazardous substances from a disposal site, a knowledge of the dynamic behavior of the groundwater flow is essential. The present paper deals with a simulation study using a finite element based Groundwater Flow Realization (GFR) program to understand the dynamic behavior of groundwater flow in the region of Schoneiche disposal site. The model may find application in the proper management of the groundwater resources of the region under consideration.

Keywords: Modeling, waste disposal site, pollutant, groundwater

Introduction

Because of extensive industrialization more than 300 million tones of waste are annually produced in Federal Republic of Germany (Knodel et al., 1993). These waste are being deposited at several thousands of disposal sites. Each waste disposal site, wheather in operation or abandoned, is within regional groundwater flow system. As a result hazardous substances seeping from a disposal site contaminate surrounding fresh groundwater regimes and surface water bodies, which are the main sources of water supply for domestic and industrial uses. According to an estimate more than fifty thousand disposable sites have been identified as sources of pollution.

The Schoneiche waste disposal site is located in south of a small town Mittenwalde, about 20 km away from Berlin. It consists of two deposits: Schoneicher Plan Deposit (SPD) and Schoneicher Deposits (SD) surrounded by Notte canal from NE-SW direction, Galluner canal from NNW-SSE direction and Schoneicher from south direction (Fig.1). The total surface area covered by both deposits is 210 ha (2.1 sq. km). The total area under investigation is about 8.5 sq. km. with 3.5 km maximum length in N-S direction and 3 km width in E-W direction. Locations of both deposits within the simulated boundary of the study area is shown in Fig. 1.

In general maximum heights of SPD and SD are 52 m and 40 m, respectively. Average elevation of the northern region is 36.5 m and of the remaining region is 37.5 m (Fig.2). In this study heights are measured with respect to mean sea level (msl), positive value for above msl and negative value for below msl. On the basis of inerpretation of geoelectrical sounding data along 8 profiles within the study area, three layers of subsurface geological formations have been delineated. These are: top layer of Quarternary sands, middle layer of quarternary silt and clay, intercalated quarternary sand and tertiary sand and silt. The bottom layer is of Tertiary sand (Hendriks and Partzsch, 1994). Bases of both depoits are more or less on top of layer 2 at an elevation of 20 m. Elevation of the top of layer 2 varies between msl in SW corner to 20 m in the entire region except in the corner of SE where the elevation is about 15 m (Fig.3). Elevation of top surface of layer 3 varies between -7.5 to -25 m (Fig.4) and its bottom surface from -80 m in SW, -50 m below SPD to -70 m in north(Fig.5). Depth of water table from the ground surface varies between 0.5 to 1.5 m.General trend of groundwater flow is from south to north and NE directions. On the basis of investigation of observed

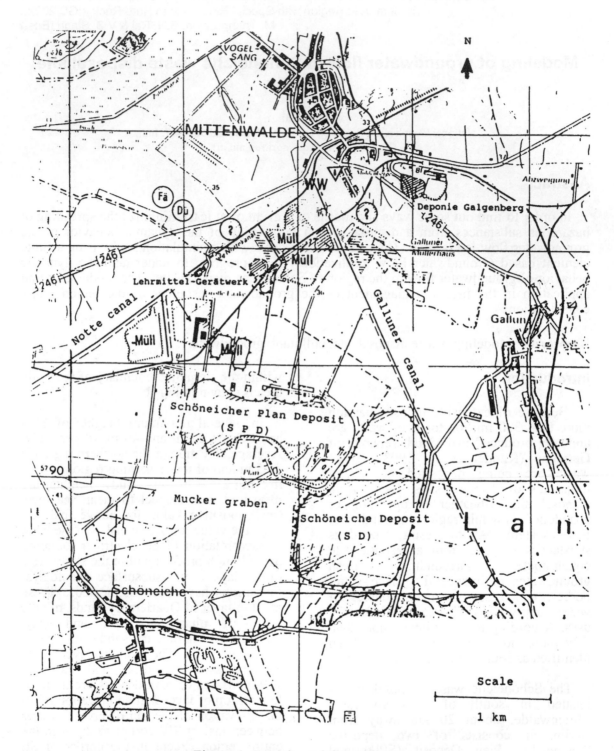

Figure 1: Location of the study area

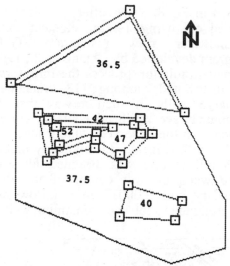

Figure 2: Elevation of top of layer 1

water table from December 1992 to October 1993, it is found that the maximum variation in water table height is 0.25 m. It shows that the groundwater flow condition is almost in steady state

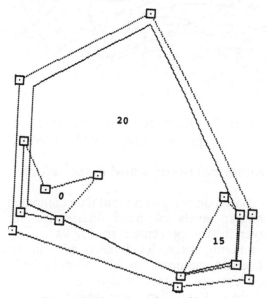

Figure 3: Elevation of top of layer 2

In the absence of detail information about spatial distribution of hydraulic conductivity value, a single value of 0.0001 m/s is considered for both horizontal and vertical hydraulic conductivities in the entire region of layers 1 and 3. Because of the presence of clay and silt material in the second layer, its horizontal and vertical hydraulic conductivity values are taken as 0.000001

m/s. On guess basis these values are taken from Freeze and Cherry (1979). Because of similar reason a single values of 2.5 X 10^{-9} m/s (=80 mm/year) for the rate of recharge due to precipitation is assigned to the entire model ara (Chiang and Kinzelbach, 1993). Annual precipitation in the study area varies between 500 to 650 mm.

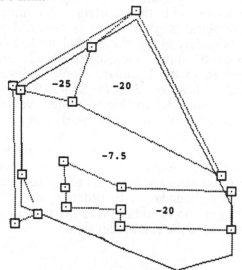

Figure 4: Elevation of top of layer 3

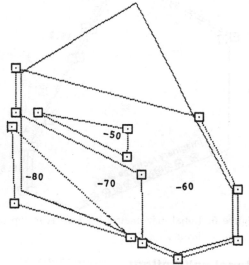

Figure 5: Elevation of bottom of layer 3

Boundary conditions

Due to general trend of groundwater flow from south to north and NE directions, southern boundary of the model area is considered as a inflow boundary. This condition is simulated by introducing 13 imaginary recharge wells at the top of each layer along this boundary (Fig.6).

149

Since the inflow rate across this boundary is not known, firstly water balance calculation for this region is carried out. Based on the water balance calculation an inflow rate of 0.0018 m³/s is considered for each layer. Boundaries along Notte canal and Galluner canal are considered as fixed head boundaries with the assumption that these canals are hydraulically connected with the aquifers. An initial piezometric head of 35 m is assigned to the boundary along Notte canal while five values of initial piezometric heads are assigned to the boundary along Galluner canal. These are 35.5 m, 35.4, 35.35 m, 35.2 m and 35.1 m (Fig.6). These values are based on the observed heads along the respective boundaries. The same fixed head areas with corresponding numerical values are assigned to all four surfaces. Rest boundaries are automatically taken as no flow boundary. In order to start the iteration. 35 m initial piezometric head is assigned to the rest model area.

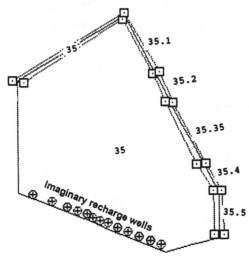

Figure 6: Initial piezometric head areas in top of layer 1

Model calibration

For verification of model results we have choosen water table contour levels of October, 1993 as shown in fig.7, because it represents for the average flow condition of the model area. After applying the above mentioned input data, the model area is descretised into small elements by defining 30 nodes in each x and y directions. Distributions of elements are

shown in Fig.8. Thereafter, simulation is carried out using a finite element based groundwater flow realization (GFR) program developed by Chiang et al., 1992. During simulation process the inflow rate is distributed among 13 imaginary recharge wells in such a way that the computed level of head contours closely match with the observed contour levels on 22-10-1992 (Fig.7). In this study 0.0001 m³/s inflow rate was assigned for each three wells on left corner of the flow boundary and 0.00015 m³/s inflow rate was assigned to the each remaining wells.

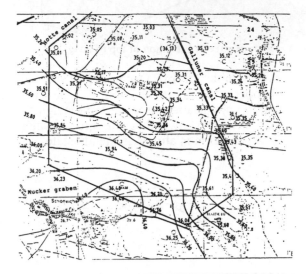

Figure 7: Observation contour levels of phreatic heads on 22-10-1993

Numerical results and discussions

Based on the given input data computed contour levels of head distribution for layer 1 are presented in Fig.9a. By comparing water level contours with the observed contours on 22-10-1993 (Fig.7) one can see a good match between both results. It confirms the validity of the present groundwater flow model for Schoneiche waste disposal site. Contour levels of piezometric head for layers 2 and 3 are given in figs.9b and c, respectively. Pattern of contour levels for layers 1 and 2 are almost similar but in case of layer 3 the contour of 36.4 m is absent.

Computed head variations in layer 1 along profiles AA', BB', and CC' (Fig.8) are presented in Figs. 10 a, b and c,

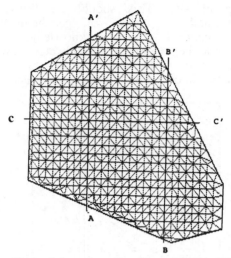

Figure 8: Locations of profiles AA', BB', CC'

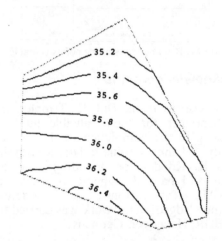

Figure 9a: Contours of phreatic heads in layer 1

upto the base of layer 1. Thereafter, the flow is vertically downward in layer 2. At the base of layer 2 the flow becomes more or less horizontal and gradually enters into layer 3. From point E the movement of groundwater through SD is in SW-NE direction and the pattern of pathline is similar to that of from point D. D' and E' are the end points of pathlines at the model boundaries for starting points D and E, respectively.

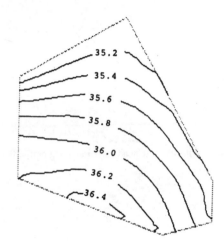

Figure 9b: Contours of piezometric heads in layer 2

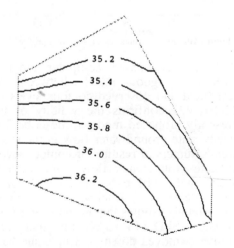

Figure 9c: Contours of piezometric heads in layer 3

respectively. Profiles AA' and BB' traverse SPD and SD from south to north direction while profile CC' traverses SPD from west to east. In these figures horizontal axes represent distances measured from the starting points A, B and C, and vertical axes represent head variations at corresponding distances along the profile. Figures 10a and b show almost linearly decline of head from south to north. But fig.10c show almost no change in head upto 1 km distance from the western boundary. After that the head decreases from a level of 35.8 m to a level of 35.35 m. Figure 11 shows the pathlines of groundwater flow from D and E points. The pathline from D shows groundwater flow through SPD in S-N direction. Groundwater gradually moves downward

Conclusion

A simulation study was carried out to understand the dynamic behaviour of groundwater flow in the region of Schoneiche waste disposal site by using a

151

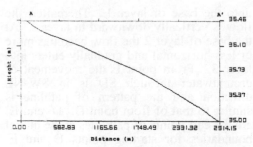

Figure 10a: Phreatic head variation along profile AA'

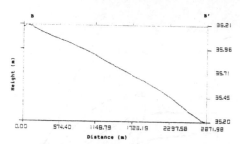

Figure 10b: Phreatic head variation along profile BB'

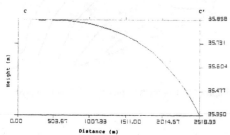

Figure 10c: Phreatic head variation along profile CC'

finite element based computer program GFR. A good match between the computed and observed contour levels on 22-10-1993 confirms the validity of the presently proposed model to describe 3-D steady state groundwater flow in the study area. Numerical results for contour levels indicate groundwater flow from south to north and north-east direction through these deposits. Analysis of pathlines direction also shows that groundwater flow through SPD in S-N direction and gradually moves downward upto the base of layer 1. Thereafter, the flow is vertically downward in layer 2. At the base of layer 2 the flow becomes more or less horizontal and gradually enters into layer 3. The movement of groundwater through SD is in SW-NE direction and the pattern of pathline is similar to that of from point D. It indicates that the groundwater in layer 1 in north of deposits is not affected by flow through waste

disposal site. This should be further confirmed from geochemical investigations. Such a knowledge of dynamic behaviour of groundwater flow can provide informations necessary for withdrawal of pollution free groundwater.

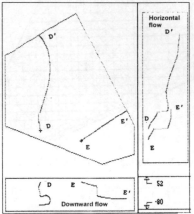

Figure 11: Pathlines for starting points D and E

Acknowledgements

I wish to thank Drs. K. Trippler, and M.Schreiner for their valuable guidance leading to the completion of this present study. Thanks are due to Drs. K.Kreysing and K. Knodel for providing necessary data. Financial support provided by DAAD at Bonn is gratefully acknowledged. This work was carried out at BGR, Hannover, Germany.

References

Chiang, W.H., Cordes. C., Kinzelbach, W. and Fang, S.Z., 1992. Groundwater flow realization (GFR), Unpubl. Report, Kassel Uni., pp.49.

Chiang, W.H. and Kinzelbach, W., 1993. Processing Modflow. Version 3.0.

Freeze, R.A.and Cherry. J.A., 1979. Groundwater. Prentice-Hall, pp. 604.

Knodel, K., Kreysing, K. and Dumke, I., 1990. Methods for studying the subsurface of planned, operating and abandoned waste disposal sites. Unpubl. Report, pp.8.

Hendriks, F. and Partzsch, K., 1994. Einschatzung der Umweltbeeinflussungen im umfeld der Deponien Schoneiche und Schoneicher Plan. Unpubl. Report, pp 48.

Intl. Conf. on Sustainable Development and Management of Groundwater Resources
in Semi-Arid Region with Special Reference to Hard Rock, (IGC 2002),
M. Thangarajan, S.N. Rai & V.S. Singh (Eds.)

An Integrated approach to groundwater resource assessment

K. Sridharan[1], P. Manavalan[2] and M. Ramanjaneyulu[3]

[1]Department of Civil Engineering, Indian Institute of Science, Bangalore, India,
e-mail: kalmb@civil.iisc.ernet.in
[2]Regional Remote Sensing Service Centre (ISRO), Bangalore, India
[3]National Remote Sensing Agency (ISRO), Hyderabad, India

Abstract

The Indian Institute of Science has been active in ground water modeling over the years, with particular focus on hard rock areas. The Indian Space Research Organization has been associated with ground water assessment and preparation of ground water potential maps for different regions in India. Recently, a joint programme has been launched to integrate the approaches of these two organizations, with a view to develop an application software which can be used by the state agencies for ground water assessment on watershed basis. The project is oriented so as to be in tune with the future strategy for ground water assessment, as indicated in the report of the Ground Water Resource Estimation Committee 1997.

The ground water assessment model integrates the following components: a) data input from satellite as well as topographic maps and other sources, b) data synthesis through image processing and GIS, c) ground water simulation through finite element model and d) parameter estimation by weighted least squares algorithm. The data structure comprises of multi-layer discretisation with different patterns of piecewise homogeneous zones for transmissivity, specific yield, recharge factor, rainfall input, draft input and administrative blocks for planning. The ground water simulation model itself is based on finite element discretisation which is overlaid on the different zones. The zonation for transmissivity, specific yield and recharge factor are arrived at based on interpreted thematic maps after assigning different weights for each thematic layer and also for each sub-units of the layer, for a given parameter.

Key words: Modeling; GIS, remote sensing, finite element model, parameter estimation, recharge

Introduction

The report of the Ground Water Resource Estimation Committee (1997) has stipulated that all hard rock regions of the country should shift to ground water assessment based on watershed as the unit for assessment. It has also recommended that in future, attempts should be made to utilize remote sensed data along with numerical modeling in regional ground water assessment. In general, watershed areas in hard rock regions may be treated to be bounded by no flow boundary with respect to ground water movement, except at the exit of the stream from the watershed. This feature facilitates numerical modeling, particularly when finite element model is used, for which no flow boundary condition forms a natural boundary condition.

Parameter estimation is a core problem dealt with by a number of investigators in ground water starting from the classic paper by Theis (1935). A large number of analytical solutions have been developed for various conceptual models of well field situations. These include confined, unconfined and semi-confined aquifer systems and double porosity systems, involving various degrees of complexities (Zekai Sen,1995; Vedat Batu,1998). These models aim to characterize the aquifer in the local well field environment.

The objective of parameter estimation models in regional ground water situation is distinctly different, in the sense that estimation of recharge is an important objective in regional ground water assessment. However, in the parameter estimation approach through ground water modeling, estimation of recharge is intrinsically linked to simultaneous estimation of the aquifer characteristics, namely the transmissivity and storage parameters. While ground water literature is replete with references on parameter estimation for characterizing the aquifer in terms of transmissivity and storage parameters, simultaneous estimation of these parameters, along with recharge in a regional ground water setting, are relatively scarce. In the Indian context with nearly 70% of land area being of hard rock terrain with relatively low recharge in terms of annual rainfall, reliable estimation of recharge becomes an important aspect for planning ground water development. A reliable assessment of recharge and a development policy in tune with such an assessment will minimize the risk of irreversible decline in ground water levels with associated scarcity conditions. Indeed, a reliable assessment of annual recharge vis-à-vis annual draft will be helpful in prioritizing water harvesting measures involving watershed development techniques.

Regional groundwater assessment

Case studies of parameter estimation on regional ground water scale have been made by Neuman et al (1980), Cooley (1983), Careera and Neuman (1986), Condon et al (1993) and Wendy and Christiana (1994). Parameter estimation models may be based on equation error criterion or output error criterion and among the two approaches, output error criterion has been the more commonly used. The Indian Institute of Science (IISc) has, over the years, used the output error approach of parameter estimation for both well field and regional ground water situations (Sridharan et al,1986; Sridharan et al,1987; Sekar et al,1994). Nagaraj (1999) conducted an extensive study on parameter estimation in regional ground water situation using the weighted least squares method in association with the finite element model for ground water simulation. He developed the techniques based on exhaustive numerical experimentation on synthetic aquifer problems and then applied them to two real life situations. These studies clearly confirm the robustness of the weighted

least squares method for the simultaneous estimation of transmissivity, specific yield and recharge factor in an unconfined aquifer system.

The Indian Space Research Organisation (ISRO) has taken up several case studies of determining ground water potential in different parts of the country. In these studies, ISRO approach involves a synthesis of information from remote sensed data, topographic sheets and ground data and characterising the aquifer and recharge potential based on lithology, structures, drainage pattern, geomorphology, soil, land use etc. The inferences made from such a composite analysis of data is verified through select field information on the yield from wells in different sub-domains of the region, before preparing a ground water potential map for the region. In this process involving synthesis of data from different thematic maps and other non-graphical data, Geographical Information System (GIS) plays a very important role.

Recently, a joint IISc-ISRO research program has been taken up to integrate the two approaches, with a view to develop an application software which can be used by the state agencies for ground water assessment on a regional scale, with watershed as a unit and with specific focus on application for hard rock areas. The following sections present an overview of this integrated approach, the issues involved in modeling, and the status of model and software development.

Conceptual Model

The parameter estimation model for ground water assessment presented in this paper is based on the unconfined aquifer system. Fig. 1 presents a schematic diagram of the unconfined aquifer system. While in general, an anisotropic system may be considered, in practice, the restriction of data availability usually may not justify the use of anisotropic transmissivities. The ground water simulation model used in this study does allow the use of anisotropic transmissivities, but in practice its application is made with isotropic transmissivity assumption. The governing equation for a two dimensional isotropic, non-homogeneous unconfined aquifer system is given by

$$\frac{\partial}{\partial x}\left(T\frac{\partial h}{\partial x}\right)+\frac{\partial}{\partial y}\left(T\frac{\partial h}{\partial y}\right)=S_y\frac{\partial h}{\partial t}+Q_g-Q_r \quad (1)$$

where h is the head or ground water level, T is the transmissivity, S_y is the specific yield, Q_g is the net pumping rate per unit area, Q_r is the recharge rate from rainfall and other sources per unit area, x and y are the Cartesian coordinates in plan and t is the time. In the above equation, h is a function of (x,y,t), T and S_y are functions of (x,y) and Q_g and Q_r are functions of (x,y,t).

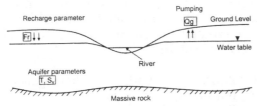

Figure 1: Schematic Diagram of Unconfined Aquifer System

Along with aquifer parameters, the present study is concerned with the simultaneous estimation of recharge parameter. For this, the recharge from rainfall only is considered. The approach may be generalized to include recharge from other sources, either as an additional known source term or as an additional term involving unknown parameters. Rainfall-recharge relationship may be governed by a complex nonlinear process when recharge from individual storm events are considered. However, when the objective is to estimate the seasonal recharge, a simple linear relationship between rainfall and recharge may be used. Rigorous one-dimensional numerical modeling of the unsaturated domain with simulation for a number of years, considering the detailed rainfall pattern, has shown that when the concern is annual or seasonal recharge, a linear rainfall - recharge relationship may be applicable (Thota Sridhar,1999). Such a simple relationship may not hold good if the rainfall in an year is very scanty or very excessive. In the present study, a simple

linear relationship is used between rainfall and recharge as follows.

$$Q_r = F_r R_f \quad (2)$$

where F_r is the recharge factor for rainfall and R_f is the rainfall rate. It is to be again noted that the use of eqn. (2) may be justified only when the objective is to obtain seasonal recharge. The merit of eqn. (2) is that it involves only one recharge parameter.

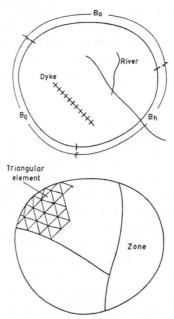

Figure 2: Schematic Diagram Showing

The solution of equation (1) requires specification of initial and boundary conditions. The known water levels at the start of the simulation period provide the initial condition. For a watershed domain, the no flow boundary condition may be applied on the external boundary. However, the simulation model allows for specification of more general boundary conditions as schematically represented in Fig. 2. The B_o boundary segment represents the no flow boundary condition, the B_q boundary segment represents specified flux boundary condition and the B_h boundary segment represents specified head boundary condition. In addition, there may be interior boundaries such as a river or a dyke as shown in Fig. 2. Along a river, the river water level may be taken as the aquifer head, treating the aquifer and the river to be a single continuum. If

155

interior dykes are present, no flow boundary condition is to be specified across the dyke, which allows for discontinuity of water level across the dyke. Such an interior no flow boundary condition is conveniently handled in the finite element method.

The present formulation is based on piecewise homogeneous system, that is the aquifer parameters or the data such as rainfall and draft are treated to be constant within a sub-domain or zone (Fig. 2) and the variability is accounted for by variations between the zones. However, the zonation pattern may vary from parameter to parameter or for data input. In this respect, the present formulation differs distinctly from that of Nagaraj (1999), in which a stationary zonation pattern is used for the transmissivity, specific yield, recharge factor, rainfall and draft. With a variable zonation pattern, the number of zones for each parameter or data input may also vary. For example, if in a watershed, if data for only one rain gauge is available, the entire watershed may be a single zone for rainfall input, but may have multiple zones for parameters such as transmissivity, specific yield and recharge factor. In the present study, the ground water simulation is done using the finite element method with triangular linear elements (Fig. 2).

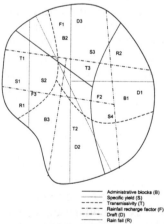

Figure 3: Schematic diagram of multiplayer discretisation

The parameter estimation problem in the regional ground water system involves the determination of the transmissivity, specific yield and recharge factor for the piecewise homogeneous zones. The total number of parameters to be estimated depends on the number of piecewise homogeneous zones for the three parameters. Rainfall and draft data are to be provided as input. Besides, water level data at some observation wells are also required. Such data may be procured at different time intervals in the period under consideration. For proper estimation of the parameters, an adequate redundancy in data is required, as real life data will invariably contain some noise.

Discretisation Structure

As stated in the previous section, the region is divided into a number of piecewise homogeneous zones, with the zonation pattern being different for different parameters and data. Specifically, the data structure comprises of multi-layer discretisation based on the following six zonation pattern:

- Zones with homogeneous transmissivity values;
- Zones with homogeneous specific yield values;
- Zones with homogeneous recharge factor values;
- Zones with homogeneous rainfall values;
- Zones with homogeneous draft values;
- Zones corresponding to administrative blocks/micro-watersheds for output processing.

Fig. 3 presents a schematic diagram of the proposed multi-layer discretisation. It may be noted that for each item above, the number of zones as well as the pattern of zonation may vary. The last zonation pattern based on administrative blocks or micro-watersheds is provided, as the development plan may be based on these units. In addition to these six zonation type discretisation, the finite element discretisation is also superimposed. Hence, essentially there are seven different layers of discretisation in the proposed model.

In the setting of such multi-layer discretisation, GIS provides a very useful tool for data processing. Essentially, in the computational module, the zonal level information is converted to the element level information and in this process, weighted averaging may be required if an element falls in more than one zone. The elements are at a finer discretisation level than a zone (the number of zones for a parameter will be far less than the number of elements), but it is possible that a zonal boundary may cut through an element and it is in such situations, when weighted averaging may be required. Time dependent data, such as rainfall and draft, are available for the respective zones and these data may have its own time series. The model is developed to allow any uniform or non-uniform interval based data for such variables.

Parameter Zonation

The zonation pattern for the parameters, transmissivity, specific yield and recharge factor is decided based on thematic maps prepared by ISRO, based on satellite data as well as other data such as from topo-sheets. At present, the model utilizes the following thematic maps, but the software is developed such that additional maps can be incorporated in the algorithm based on future availability. The thematic maps considered now are: a) lithology, b) lineaments, c) geomorphology (for land form), d) drainage pattern and e) irrigated area (for land use).

The principle underlying the zonation discretisation is as follows. For each of the three parameters, namely transmissivity, specific yield and recharge factor, a certain weight is associated for the attribute based on each thematic layer. For example, the lithology map may have a larger weightage attached to transmissivity, while land use may have a larger weightage attached to recharge factor. Based on such a weightage scheme, a numerical index is developed as follows.

$$F(p) = \sum w_i(p) x_i \qquad (3)$$

where F(p) is the numerical index associated with the parameter p in a sub-polygon, $w_i(p)$ is the weight associated with the ith thematic layer (which may be either polygon or line coverage) for the pth parameter and x_i is the numerical value

associated with the attribute of the sub-polygon. The numerical index F(p) is computed in all the sub-polygons formed out of the merger of all the polygons from the different thematic maps. This is illustrated in Fig. 4 showing the combined polygon map from different thematic layers. The data from line coverage is also evaluated in these sub-polygons.

Figure 4: Schematic diagram illustration merged output of different layers

Once the values of F(p) are determined for all the sub-polygons, the zonation pattern for each parameter p is finalized based on apportioning F(p) into the number of zones specified by the user, forthe particular parameter p. The number of zones may be decided based on a knowledge of hydrogeology of the region as well as limitations of data availability. As stated earlier, a certain amount of redundancy in the data is necessary for reliable estimation of the parameters.

In applying the above principle for determining the zones for the different parameters, specification of the numerical values of $w_i(p)$ and x_i is necessary. At present, this is done heuristically, but the specification of these values is best done by wide consultation among hydrogeologists based on their knowledge of the terrain for different regions in the country, particularly in hard rock area. It may, however, be noted that these values are stationary inputs in the program and can be improved at any time, based on the best current estimates.

WLS Algorithm

The parameter estimation of regional ground water system is mathematically modeled as a nonlinear programming problem. The objective is to estimate the set of unknown parameters, $T(i)$, $S_y(j)$ and $F_r(k)$, $i = 1$ to Z_i, $j = 1$ to Z_j and $k = 1$ to Z_k, where Z_i refers to the number of transmissivity zones, Z_j refers to the number of specific yield zones and Z_k refers to the number of recharge factor zones. Thus, in all, $Z_i + Z_j + Z_k$ parameters are to be determined. These parameters are to be estimated such that the computed values of the water levels match well with the observed values in the field. The weighted least squares algorithm for parameter estimation is one of the most commonly used approaches for the solution of such inverse problems in ground water literature. This approach minimizes the sum of squares of the differences between computed and observed values.

The objective function is expressed by the following equation,

$$E = \sum_{L=1}^{NS} \sum_{i=1}^{M(L)} W_i^L \left[s_{oi}^L - s_{ci}^L \right]^2 \qquad (4)$$

where NS is the number of seasons or time intervals for which measurements are

made, s_{oi}^L is the observed value of variable i for season L, s_{ci}^L is the computed value of variable i for season L, M(L) is the number of measurements in season L and W_i^L is the weight for i^{th} measurement in season L. The term "season" is used in a general sense here reflecting measurements in several time intervals in a year. The time intervals may or may not be uniform. For example, water level measurements at observation wells may be made every month or only for pre and post-monsoon cut-off times.

In the present study, the Gauss-Newton algorithm is used for the minimization of the objective function as it has been demonstrated to be very robust and reliable (Nagaraj, 1999). If N is the total number of parameters to be estimated ($Z_i + Z_j + Z_k$), the simultaneous linear equations for parameter corrections p_l (l=1 to N) in any outer or parameter iteration are derived based on the first order Taylor series approach. Using the Taylor series expansion, the expression for s_{ci} can be written as

$$s_{ci}^L = s_{*i}^L + \left[\sum_{i=1}^{N} \left(\frac{\partial s_i^L}{\partial p_{*l}} \right) \Delta p_l \right] \qquad (5)$$

where s_{*i}^L is the computed value of s_i^L corresponding to the set of parameters, p_{*l}, the initial trial values of the parameters or the values obtained from the previous iteration, p_l is the respective increment to p_l in the current iteration to yield improved parameter estimates, s_{ci}^L is the improved (computed) value of the variable and $\partial s_i^L / \partial p_{*l}$ is the sensitivity coefficient of the ith measurement in season L with respect to the parameter p_l. The sensitivity coefficients are evaluated based on the parameter values from the previous iteration using a forward difference approximation. This requires simulation for (N+1) cases to determine all the required sensitivity coefficients.

Using equation (5) in equation (4) and equating the partial derivatives of E with respect to p_l to zero, the following set of linear equations for the parameter corrections are obtained.

158

$$\sum_{m=1}^{N} \left[\sum_{L=1}^{NS} \sum_{i=1}^{M(L)} W_i^L \frac{\partial s_i^L}{\partial p_{*l}} \frac{\partial s_i^L}{\partial p_{*m}} \right] \Delta p_i$$

$$= \sum_{L=1}^{NS} \sum_{i=1}^{M(L)} W_i^L \frac{\partial s_i^L}{\partial p_{*l}} \left[s_{oi}^L - s_{*i}^L \right]$$

$$1 = 1....N \qquad (6)$$

The solution of eqn. (6) yields the parameter corrections for updating the parameters as follows.

$$p_l^{k+1} = p_l^k + \Delta p_l^k \qquad 1 = 1....N \qquad (7)$$

In applying eqn. (7), an additional condition on limits to corrections is stipulated based on several earlier studies (Nagaraj, 1999) to ensure convergence of the parameter iteration process.

$$-0.2 \leq \frac{\Delta p_l}{p_l} \leq 0.5 \qquad (8)$$

Choice of Weights

The choice of weights in the weighted least squares algorithm is based on the conclusions obtained from the extensive studies of Nagaraj (1999) for both synthetic and real aquifers. Nagaraj (1999) has recommended two approaches for the choice of weights in the weighted least squares method. Where there is good data redundancy, a two step procedure is proposed, with mean measured values as weights in the first step and information from measurement covariance matrix obtained from the first step used for deciding the weights in the second step. Alternately, where the data redundancy is not good, a sequential estimation procedure is proposed. In this procedure, the transmissivities are estimated in the first cycle with the specific yield and recharge factors treated as known from assumed or previous iteration values. In the next cycle, the specific yield and recharge factor values are estimated retaining the transmissivity values obtained from the previous cycle. The process is repeated until all the parameters converge. In the sequential estimation procedure, each cycle involves solution of the parameter estimation problem. Hence the limitation of data is partially overcome at the cost of additional computational time. Nagaraj (1999) has demonstrated the efficacy of the method for real aquifer systems.

Role of GIS

The entire software for the solution of the parameter estimation problem is embedded in GIS environment with the main or control program written in Arc Macro Language (AML). GIS comes in very handy in managing the multi-layer discretisation structure of the model. The preliminary data processing is done using GIS and the Fortran program for ground water simulation and parameter estimation is called for the detailed computations. The post-processing of the results is also done using GIS. Specifically, the following data processing steps are done through GIS.

- Conversion of real world maps into digital maps by digitization or scanning;
- Processing of thematic layer maps obtained from ISRO for determining the zonation pattern for the parameters;
- Generation of finite element mesh and establishing element connectivity;
- Mapping element-zone relationship for calculation of parameter values in the elements;
- Prescribing initial nodal water levels from ground water contours;
- Mapping element-zone relationship for providing input of rainfall or draft data for a particular time step;
- Mapping the results of analysis from element scale to administrative block or micro-watershed scale;
- Secondary computations like areas of elements, zones etc.

A simulation option is also provided in the software for analysis of a situation in which the parameters are already established and the effect of a particular development policy is to be studied. In this option, the global water balance quantities for the entire region and separately for the administrative blocks in the region are provided.

The software is developed and tested in two phases, the ground water simulation and parameter estimation program along with the GIS interface tested separately. The data processing from the thematic layers provided by ISRO and the establishment of the multi-layer zonation pattern based on the analysis of these data is developed and tested separately. While the former process is completed, in the latter process, the assignment of numerical values to the attributes from the different thematic layers and the associated weights for the three types of parameters, namely transmissivity, specific yield and recharge factor, as presently done is only a first approximation. It is proposed to improve on this significantly based on extensive consultation among hydrogeologists, with specific focus on hard rock areas.

Conclusion

An integrated approach for regional ground water assessment is proposed, based on the use of thematic maps from remote sensed data and other data, along with finite element model for ground water simulation and weighted least squares method for parameter estimation. While the principal objective of the assessment is the estimation of seasonal or annual recharge, the process necessarily involves simultaneous estimation of the aquifer parameters, namely transmissivity and specific yield. A multi-layer discretisation is proposed with distinct piecewise homogeneous zonation pattern for the three parameters, transmissivity, specific yield and recharge factor, and the two data inputs, namely rainfall and draft. Besides, the administrative blocks are also used as separate zonation for facilitating development plans. The zonation pattern for the three parameters are based on an assessment of the effect of the attributes in different thematic layers, consisting of both polygon and line features. These processes are greatly facilitated by the use of GIS and hence the entire software is developed in GIS environment.

Acknowledgement

The work reported in this paper benefitted considerably form the inputs provided by a review committee formed by ISRO. The writers are thankful to the members of the committee, Mr. S Adiga, Dr. P P Nageshwara Rao, Dr. P R Reddy, Mr. V Tamilarasan and Mr. A K Chakrborty for their valuable suggestions.

References

Careera, J. and Neuman, S.P., 1986. Estimation of aquifer parameters under transient and steady state conditions. Water Resources Research, 22(2), 211-227.

Condon, M.R., Traver, R.G., Fergusson, W.B. and Chadderton, R.A., 1993. Parameter estimates for a groundwater model. Water Resources Bulletin, 29(1), 95-106.

Cooley, R.L., 1983. Incorporation of prior information on parameters into non linear regression groundwater flow models – Applications. Water Resources Research, 19(3), 662-676.

Ground Water Resource Estimation Committee Report, 1997. Ministry of Water Resources, Government of India.

Nagaraj, M. K., 1999. Parameter estimation of regional groundwater systems. Ph D thesis, Indian Institute of Science, Bangalore.

Neuman, S.P., Fogg, G.E. and Jacobson, E.A. 1980. A statistical approach to the inverse problem in aquifer hydrology – a case study. Water Resources Research, 16(1), 33-58.

Sekhar, M., Mohan Kumar, M.S. and Sridharan, K., 1994. Parameter estimation in an anisotropic leaky aquifer system. J. Hydrol., 163(4), 373-391.

Sridharan, K., Lakshmana Rao, N.S., Mohan Kumar, M.S. and Ramesam, V., 1986. Computer model for Vedavati ground water basin : Part 2 – Regional Model. Sadhana, Proc. Indian Academy of Sciences, 9, Part 1, 43-55.

Sridharan, K., Ramaswamy, R. and Lakshmana Rao, N.S. 1987. Identification of parameters in semiconfined aquifers. J. Hydrol. 93, 163-173.

Theis, C.V., 1935. The relation between lowering piezometric surface and rate and duration of discharge of a well using groundwater storage. Trans. Am. Geoph. Union, Part 2, 513-534.

Thotha Sridhar, 1999. Unsaturated flow modelling for ground water recharge assessment. M.Sc [Engg] thesis, Indian Institute of Science, Bangalore.

Vedat Batu, 1998. Aquifer Hydraulics, John Wiley & Sons Inc.

Wendy, D.G. and Christiana, R.N. 1994. Optimal estimation of spatially variable recharge and transmissivity field under steady state groundwater flow – Case study. J. Hydrol. 157, 267-285.

Zekai Sen, 1995. Applied Hydrogeology, CRC Press, Lewis Publishers.

Intl. Conf. on Sustainable Development and Management of Groundwater Resources
in Semi-Arid Region with Special Reference to Hard Rock, (IGC 2002),
M. Thangarajan, S.N. Rai & V.S. Singh (Eds.)

Use of cross-correlation and spectral analysis to characterize the groundwater level fluctuation of a hard-rock aquifer

J.C. Marechal[1] and K. Subrahmanyam[2]

[1] Bureau de Recherches Geologiques et Minières (French Geological Survey) and
[2] National Geophysical Research Institute, Indo-French Centre for Groundwater Research, Hyderabad, India

Abstract

The groundwater level fluctuations constitute mainly the response of the aquifer to the stress induced by rainfall during monsoon. It is therefore useful to characterise this response in order to define the hydraulic behaviour of the aquifer. The cross-correlation and spectral analysis techniques are applied on rainfall data and groundwater level fluctuations from 11 observation wells located in a hard-rock region of Andhra Pradesh, near Hyderabad. These preliminary results show the primordial influence of water-table depth on the groundwater level fluctuations, namely after the rainfall. These techniques are highly useful for the location of an artificial recharge structure.

Key words : Water level fluctuations, hard rock aquifer, groundwater, cross-correlation, spectral analysis, recharge

Introduction

The use of time series data can give interesting information on hydrologic process and impulse response characteristics of aquifer systems. Until now, most of the studies of the hydrogeological processes using correlation and spectral analyses have been focused on karstic systems. The data mainly used in the literature were rainfall as an input and discharge as an output. Laroque et al. (1998) analysed the cross-correlation and the cross-spectrum using precipitation as input and piezometric head as output. As stated by these authors, although the water level (piezometric head) fluctuations are not, physically, outputs of the aquifer system, the cross-correlation between the site rainfall and the corresponding water level variations can reveal useful information on the state of the aquifer system and on the dynamics of the water-level changes. The study of Jin-Yong Lee and Kang-Kun Lee (2000) is the only one applied to groundwater level fluctuations in fractured systems and this paper is largely inspired by their study.

Monsoon period is very important for the aquifer water balance in semi-arid areas, such as Hyderabad, India. Because discharge data from springs are not available in a typical fractured aquifer system, unlike a karstic system, we can instead analyse the easily available water-level variations, which represent the response of the fractured aquifer system to the recharge during the monsoon period. In this study, we examine the different characteristic response behaviours of the water-level variations at eleven wells in the same geologic formations.

The objective of this study is to characterise the mechanism of groundwater-level changes and to understand the groundwater recharge mechanism of a fractured basaltic/granitic aquifer through correlation and spectral analysis of hydrologic time series data.

Methods of analysis

For the mathematical expressions of the functions, please refer to Jenkins and Watts (1968), Mangin (1984), Box et al. (1994), Padilla et al. (1994) and Larocque et al. (1998). The functions include auto-correlation, spectral density and cross-correlation. The auto-correlation functions quantify the statistical dependency of successive values over a time period (Larocque et al., 1998) and memory effect (Angelini, 1997). If the time series is uncorrelated, auto-correlation function decreases very quickly and reaches a value of zero in a short time lag (bold curve - Figure 1). However, if the time series has strong interdependency and a long memory effect, the auto-correlation function shows a gently decreasing slope and nonzero values over a long time lag (Figure 1).

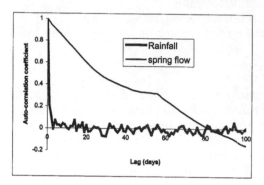

Figure 1 : Auto-correlation function of daily rainfall and of a spring discharge rate

The spectral density function is a Fourier transformation of the auto-correlation function. From the spectral density function the periodical characteristic of a time series can be identified. In figure 2a, a theoretical signal is characterised by three periods T_1, T_2 and T_3. Typically, the spectral density function of this signal shows three peaks at frequencies $f_1=1/T_1$, $f_2=1/T_2$ and $f_3=1/T_3$ (Figure 2b). The regulation time obtained from the spectral density function defines the duration of the influence of the input signal and gives an indication of the length of the impulse response of the system (Larocque et al., 1998).

The cross-correlation function represents the statistical inter-relationship between the input and the output series. If the input stress is an uncorrelated process the cross-correlation is an impulse response function of the system or aquifer (Padilla and Pulido-Bosch, 1995; Larocque et al., 1998). The delay, which is the time lag between lag = 0 and the maximum cross-correlation, determines the stress transfer velocity of the system. The cross-amplitude function identifies the filtering characteristic of the periodic components of the input stresses by the aquifer system (Padilla and Pulido-Bosch, 1995).

Study area and data characteristics

The study area is located in the Madak district, in Andhra Pradesh, 50 kilometers North of Hyderabad (Figure 3). The climate is semi-arid with an average of 800 mm/yr rainfall falling mainly between June and September. In this hard-rock region, the geology is composed of Deccan Traps and Archean granites and gneiss. The wells are about 50 meters deep and three of them are located in basalts (H, R1 and R2). The other ones are located in granite.

The time series data used in this study are groundwater-level fluctuations and site rainfall. At the origin, water levels were measured hourly by Andhra Pradesh Groundwater Department using automatic data loggers. As we focus on seasonal fluctuations, only daily data were retained. Daily rainfall data were measured at different meteorological stations, that are located near to the specific wells. The cross-correlation analysis is done only during the active period between rainfall and responding water level data (Table 1).

Results of univariate analysis

Rainfall variations

To understand characteristics of variation in site rainfall, auto-correlation functions of this variable was calculated and compared with rainfall data from an other climatic region (as an example, Abbeville from temperate climate in France). Fig. 4 shows auto-correlation function of the 10 daily rainfalls series in the study area (Table 1) and one in France. The auto-correlation coefficients of Indian rainfall decrease slowly over a long time lag. After, the correlation coefficient remains negative but not zero and increases to become positive after 300 days and reaches another maximum for k=1 year. This illustrates the annual and low rainfall during winter and summer repetition behaviour of the rainfall in India, under the influence of the monsoon. The high rainfall during one monsoon (between June to October) is correlated with high rainfall of next monsoon one year later (November to May) is anti-correlated with high rainfall with a lag of

Well Location	L	Level data starting	Level data end	Rainfall Station	Rainfall-data start	Rainfall-data end	Statistic Start	Statistic end	Statistic length (days)
Hadnur1	B	25/05/99	22/02/01	Nyalkal	01/01/99	01/01/01	25/05/99	01/01/01	587
Ranjole1	B	25/05/99	21/03/01	Zaheerabad	01/07/98	01/01/01	25/05/99	01/01/01	587
Ranjole2	B	25/05/99	24/10/00	Zaheerabad	01/07/98	01/01/01	25/05/99	24/10/00	518
Dudeda	G	26/05/99	21/03/01	Kondapak	01/07/98	11/10/00	26/05/99	11/10/00	504
Gajwel	G	22/01/98	18/08/00	Gajwel	14/06/98	12/03/01	14/06/98	18/08/00	796
Medak	G	03/01/98	05/02/00	Medak	30/06/98	02/01/01	30/06/98	05/02/00	585
Mulugu	G	24/01/98	22/03/01	Mulugu	01/07/98	13/03/01	01/07/98	13/03/01	986
Narsapur	G	06/01/98	21/03/01	Narsapur	30/06/98	10/10/00	30/06/98	10/10/00	833
Sangareddy	G	07/01/98	22/03/01	Sangareddy	01/07/98	02/01/01	01/07/98	02/01/01	916
Yeldurthy	G	26/05/99	21/03/01	Yeldurthy	01/01/99	11/10/00	26/05/99	11/10/00	504
Shankarampet	G	25/05/99	20/03/01	Shankarampet	02/01/99	29/09/00	25/05/99	29/09/00	493

Table 1: Data availability periods for each well. L: Lithology; B: Basalt; G:Granite

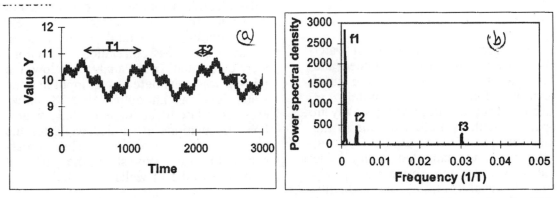

Figure 2: (a) Example of a theoretical three periods-function, and (b) its power spectral density function (T is a unit of time).

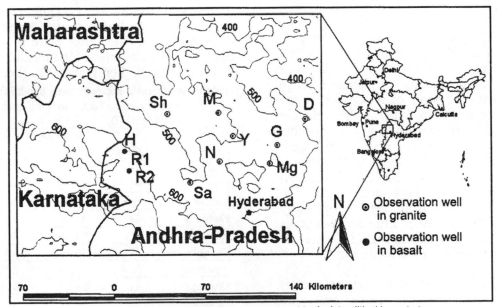

This is rough sketch only and does not purport to depict political boundaries.

Figure 3: Location map of observation wells

about half a year. In contrast, the rainfalls in France shows very different behaviour. The auto-correlation functions of the rainfalls in France quickly decrease and reach a zero value after a few days. This is an indicator of the uncorrelated characteristic of the rainfall in France. This statement is confirmed by the spectral analysis. The relative spectral density functions of Indian rainfalls show a main peak well identified at about one year (on Figure 5, see the example of rainfall at Yeldurthy with a period of about one year - 340 days), while there is no well defined main peak for this function in France rainfall.

Groundwater level fluctuations

The auto-correlation functions of water levels in wells have been computed and are compared in Figure 6 to the auto-correlation function of rainfall at Yeldurthy. They are decreasing slowly in comparison with rainfall: it is due to the memory-effect of the aquifer system (unsaturated and saturated zone). The aquifer system transforms a lowly structured signal (rainfall) into a highly structured signal (water level). The auto-correlation functions are different from one well to an other one and the delay for auto-correlation coefficient reaching r = 0 (the memory effect) is in the range from 60 to 250 days. The longest delay is observed at Mulugu well (M) while the shortest one is observed in Dudeda well (D). It is interesting to compare these delays with the depth of water level in the concerned wells. In Table 2, the wells are classified by increasing delays for r=0 that are compared to water depth in the well. The four wells with water depth superior to 20 meters (G, R1, Sa and Mg) are all characterised by delays superior than 100 days. The relationship between memory effect of the system and the water depth is illustrated on a graph in Figure 7 where an increasing trend is clearly present.

It signifies that the fluctuations of groundwater levels in these wells are strongly dependant from the water depth, thus from the thickness of the unsaturated zone. The effect of lithology can be evaluated trough the wells drilled in basalts. In Figure 7, Ranjole2 (R2) and Hadnur1 (H) wells are located in the cluster of granite wells. Only, the well of Ranjole1 (R1) behaves differently with a middle memory effect for a very deep water level. This first result indicates that at great depth, water level fluctuations in basalts could be higher than in granites. However this result should be confirmed by other observations.

Results of bivariate analysis

Cross-correlation and recharge implications

To identify a characteristic relationship between the site rainfall and the corresponding natural water-level variations, cross-correlation functions were calculated. In the analyses below, the rainfall was considered as an input and the water level as an output. Fig. 8 shows cross-correlation functions of daily water-level fluctuations at studied wells for the period of observation. The cross-correlation functions show the different responses of the 11 wells to rainfall.

The four wells already characterised above (R1, Mg, G and Sa) by long memory effect due to deep water levels are represented in bold. The maximum of their cross-correlation function are low (below 0.15) and come late (after more than 100 days). Cross-correlation functions of other wells own a higher maximum (between

Table 2: comparison between water depth in wells and memory-effect (delay for auto-correlation coefficient = 0).

Well Name	Well Code	Water depth (m)	Delay r=0 (days)
Dudeda	D	16.8	62
Narsapur	N	10.66	68
Medak	Mk	14.83	80
Hadnur1	H	13.96	81
Ranjole2	R2	14.86	90
Shankarampet	Sh	10.78	95
Yeldurthy	Y	10.78	125
Gajwel	G	26.37	128
Ranjole1	R1	40.27	132
Sangareddy	Sa	20.66	221
Mulugu	Mg	25.96	238

164

Table 3: Characteristics obtained from cross-correlation analysis (wells are classified based on increasing water level depth below ground level).

Well Location	Well code	Water depth (m)	Maximum Cross-correlation	Time delay (days)	Cross-correlation area (days)	Water-level range of fluctuation (m)
Narsapur	N	10.66	0.15	90	15.02	9.65
Shankarampet	Sh	10.78	0.23	58	27.29	10.01
Yeldurthy	Y	10.78	0.21	90	26.43	11.10
Hadnur1	H	13.96	0.28	57	29.10	11.47
Medak	M	14.83	0.23	135	25.68	5.79
Ranjole2	R2	14.86	0.23	89	24.00	7.02
Dudeda	D	16.80	0.24	88	24.81	8.32
Sangareddy	Sa	20.66	0.14	173	25.31	6.30
Mulugu	Mg	25.96	0.10	177	18.37	2.42
Gajwel	G	26.37	0.15	130	20.77	15.63
Ranjole1	R1	40.27	0.15	90	14.54	7.85

0.20 and 0.30) for a shorter delay (less than 100 days). Here again, the influence of water depth – in fact, of the unsaturated zone - is primordial on the relationship between rainfall and water level fluctuations. This is illustrated in Table 3 where characteristics of the cross-correlation functions are compared to the depth of water. The area of the positive part of the cross-correlation function indicates the amplitude of the reaction of the water-level to rainfall.

Except at Narsapur-well (N), the maximum cross-correlation and the area of the cross-correlation are high for shallow water level. Longer is the path between soil and the saturated zone through unsaturated zone, lower is the amplitude of the reaction of the aquifer. This trend is confirmed by the range of water-level fluctuations that generally decreases when the water-level increases.

Application to artificial recharge system location

The location of any artificial recharge system should integrate, among other constraints, the hydraulic capacity of the aquifer to absorb water. One way to evaluate this capacity can be found through the observation of water infiltration under natural conditions. The ideal site for infiltration system would be characterized by high stress transfer velocity and high porosity in order to

assure, respectively, high and rapid flow of infiltration and large volume of storage. On a cross-correlation function, the delay, which is the time-lag between lag = 0 and the maximum cross-correlation, determines the stress transfer velocity of the system and the surface of the cross-correlation function gives indication on the amplitude of the relation between rainfall and water levels, related to the porosity of the aquifer. Finally, an ideal site for percolation tank would be characterized by low delay and low surface of the cross-correlation function. This technique allows to determine which area is, hydraulically, well suited to welcome an efficient infiltration structure as a percolation tank for example.

On Figure 9, the studied wells are represented on a graph with the delay for maximum correlation and the area of the cross-correlation function. The best place for the location of an artificial recharge structure as a percolation tank should be near the left-bottom corner of the graph. Inversely, the worst site would be near the right-up corner of the graph. In our study area, two sites show better characteristics than others: Narsapur-Well (N) and Ranjole1 (R1) are represented by the nearest points from left-bottom corner.

Conclusion

The application of cross-correlation and spectral analysis to rainfall and

165

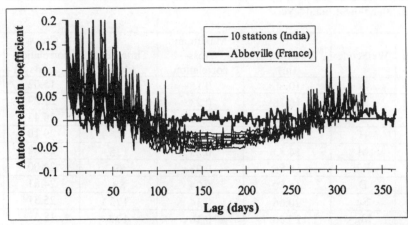

Figure 4: Auto-correlation functions of 10 rainfall data series in the study area, compared with auto-correlation function of Abbeville station in France.

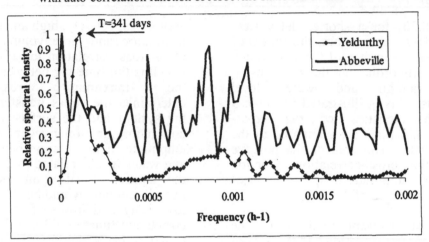

Figure 5: Relative spectral density of rainfall at Yeldurthy in the study area, compared with rainfall at Abbeville station in france.

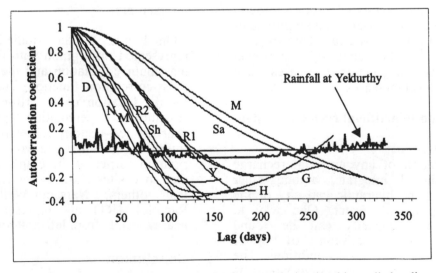

Figure 6: Auto-correlation function of daily groundwater level in studied wells.

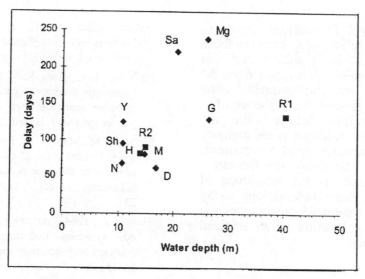

Figure 7: Relation between water depth in the wells and memory effect determined from auto-correlation function (delay for r = 0).

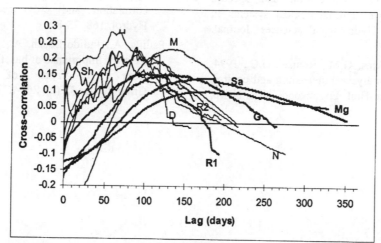

Figure 8: Cross-correlation functions of studied wells (G, R1, Sa and Mg curves in bold)

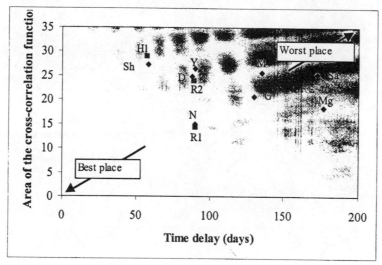

Figure 9: Representation of observation wells on cross-correlation delay – area graph

groundwater level fluctuations allows to characterise the relationship between these time data series. It is shown that this relationship is mainly dependant from the thickness of the unsaturated zone separating the ground from the aquifer. Deeper is water level, longer is the path from rain to aquifer, longer is the memory effect of groundwater level fluctuations, longer is also the delay for the stress transfer and lower is the amplitude of fluctuations. A new methodology using these signal treatment techniques is proposed for the location of an artificial recharge structure.

References

Angelini, P., 1997. Correlation and spectral analysis of two hydrogeological systems in Central Italy. Hydrological sciences. Journal. 42 (3), 425-439.

Box, G.E.P., Jenkins, G.M., Reinsel, G.C., 1994. Time series Analysis: Forecasting and control, vol. 3, Prentice-Hall, Englewood Cliffs, NJ (p. 598)

Jenkins, G.M., Watts, D.G., 1968. Spectral Analysis and its applications, Holden-Day, San Francisco, CA (p-525)

Jin-Yong Lee, Kang-Kun Lee, 2000. Use of hydrologic time series data for identification of recharge mechanism in a fractured bedrock aquifer system. J. Hydrol. 229, 190-201.

Larocque, M., Mangin, A., Razack, M., Banton, O., 1998. Contribution of correlation and spectral analyses to the reginal study of a karst aquifer (Charente, France). J. Hydrological. 205, 217-231.

Mangin, A., 1984. Pour une meilleure connaissance des systemes hydrologiques a partir des analyses correlatoire et spectrale. J. Hydrol. 67, 25-43.

Padilla, A., Pulido-Bosch, A., 1995. Study of hydrographs of karstic aquifers by means of correlation and cross-spectral analysis. J. Hydrol. 168, 73-89.

Padilla, A., Pulido-Bosch, A., Mangin, A., 1994. Relative importance of baseflow and Quickflow from hydrographs of karst springs. Groundwater 32 (2), 267-277.

GROUNDWATER POLLUTION AND ITS REMEDIAL MEASURES

Intl. Conf. on Sustainable Development and Management of Groundwater Resources
in Semi-Arid Region with Special Reference to Hard Rock, (IGC 2002),
M. Thangarajan, S.N. Rai & V.S. Singh (Eds.)

Nitrate pollution in groundwater , Krishna delta, India

V.K. Saxena, I. Radhakrishna, K.V.S.S.Krishna and V. Nagini
National Geophysical Research Institute, Hyderabad 500007
Email: vks_902001@yahoo.com

Abstract

Nitrate contents have been quantitatively estimated in the ground water samples of Krishna Delta (longitude E 80^0 10' - 81^0 15', latitude N 15^0 40' - 16^0 45'). Hydrochemical data indicate a large variation of nitrate from 5-135 mg/l. In 144 groundwater samples, about 32%have shown high nitrate contents (>50mg/l) which is more than the permissible limits of drinking water. In north Krishna delta 22% and in south Krishna delta 10% water samples were found to exceed the permissible limits. Nitrate pollution level is found more in dugwells compared to handpumps / bore wells. This study indicate that: Ground water of north Krishna delta is more polluted than south and in this region 24% dugwells and and 18% hand pumps have exceeded the desirable limits. The possible sources for the high nitrate level in ground water has been identified as excessive utilization of nitrogenous fertilizers, insectisides and pesticides for agricultural purposes.

Keywords: Nitrate pollution, groundwater, Krishna delta, fertilizers

Introduction

The presence of high nitrate concentration(>50mg/l) in ground water would normally indicate pollution of groundwater. The World Health Organization (WHO,1993) standard was originally set at 45 mg/l nitrate has been adjusted to 50 mg/l (Chettri and Smith, 1995, Canter, 1997). Since presence of excess nitrate ions in groundwater is harmful to health, their occurrence in high concentrations in groundwater is a matter of great concern. The leaching of nitrates from agriculture land has been a major research focus in past two decades. Although inorganic nitrogen fertilizers, septic tanks, poor dug wells and defective severage systems are the suspected major sources of nitrate in groundwater (Piskin, 1973, Canter, 1997, Lindsay, 1997).

The study area is composed of a delta which is known as Krishna delta. Krishna delta is located in south eastern part of India. The Krishna river and deltaic materials like alluvium, mud, peat, sand, fine grained clays and delta deposits make the Krishna delta very fertile for the agriculture point of view. The Krishna river divides the delta in two parts (1) north Krishna delta (towards the right side of the river) and (2) south Krishna delta (towards the left side of the river). Tha sand plain aquifers are the main source of water for agriculture and domestic purposes. The deltaic aquifers are recharged by direct infiltration of precipitation. The dugwells, borewells and handpumps are the major drinking water source in the Krishna delta.

As a part of research and development activity in the Krishna delta particularly for the assessment of groundwater quality, 144 groundwater samples were collected during september, 1996 (post monsoon).

The purpose of this work is to determine the extent of nitrate in groundwater in selected regions of north and south Krishna delta and to study the possible source of nitrate pollutants and suggest possible measures for the protection of groundwater quality in the Krishna delta.

Location

The regions for the study of nitrate in groundwater covers both the urban and rural areas of Krishna delta (Fig.1). In total 144 groundwater samples, 65 were collected from dugwells, 5 from bore wells and 73 from handpumps. The depth of the water in dugwells are in between 1.3 to 12 mts, whereas 7 to 50 mts in handpumps.

Health problems from Nitrate

It is well established, if nitrates are consumed more than the permissible limit (>50mg/l), it may lead to several types of diseases (Reddy, 1981, Dudley, 1990). Chemically during nitrate - water reaction, the nitrates are converted into nitrites and such reactions take place in the digestive system of human body. The nitrates oxidise the heamoglobin to methemoglobin and cause for several types of disease which mostly depend on duration and quantity of nitrite consumption (Perlistein and Attala, 1976, Dudley, 1990). The Methewmoglobin is a pigment which is incapable of acting as oxygen carrier in blood vessels. The consumption of nitrate rich water cause a large number of diseases like dizziness, abdominal disorder, vommiting, weaknesses, high rate of pulptation, mental disorder and even stomach cancer etc. (Perlistein and Attala, 1976, Reddy, 1981, and Thind, 1982, Burt and others 1993).

Methodology

The water samples were collected in double stopper 100 ml polythene bottles. Nitrate electrode, double junction reference electrode and expandable ion analyser were used to measure the nitrate concentration in groundwater (Saxena, 1991). The results of nitrate concentrations are shown in Table 1.

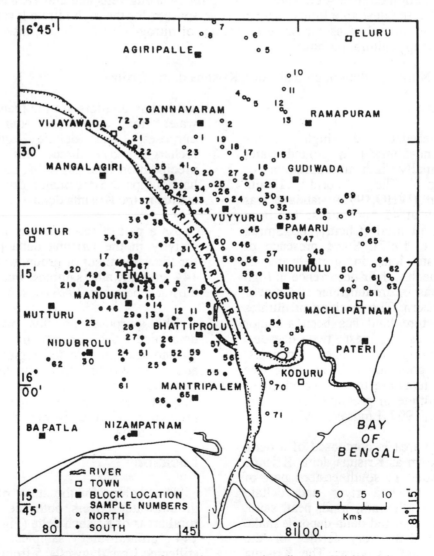

Figure 1: Location of groundwater sample collection, Krishna delta

170

Table 1: Place of sample collection,source,depth,Nitrate and pH

North Krishna Delta

S.No	Name	Type	Depth	Nitrate	pH
1	Kesoropalle	D.W	3.6	104	7.8
2	Bapeshwaram	B.W	60.0	116	7.7
3	Ambapuram	B.W	25.0	10	7.5
4	Ambapuram	D.W	6.0	50	7.5
5	Mirzapuram	B.W	15.0	10	7.6
6	Gollapalle	H.P	21	23	7.5
7	Bunivanchol	D.W	8.0	96	7.5
8	Ravicherla	D.W	3.0	64	7.6
9	Chukkapalle	D.W	8.0	50	7.8
10	Tallamudi	D.W	4.0	48	8.3
11	Kanukumulu	D.W	1.8	20	7.8
12	Aragaluru	D.W	12	54	7.5
13	Telegraph nagar	H.P	10	23	7.6
14	Gudivada	D.W	2.4	52	7.8
15	Dosapadu	D.W	2.2	10	7.8
16	Venkatpragudu	H.P	16	56	7.5
17	Indupalle	D.W	4.9	88	7.7
18	Malakadu	H.P	16	72	7.9
19	Tarogupalle	H.P	16	116	7.9
20	Kantipadu	D.W	8	40	7.5
21	Vizayawada	H.P	35	42	7.5
22	Vizayawada	D.W	4	10	10.6
23	Uppuluru	H.P	16	54	7.7
24	Vuyyuru	H.P	15	10	7.6
25	Nappalle	H.P	20	10	7.6
26	Akumuru	H.P	25	10	7.5
27	Koturu	D.W	2.9	46	7.5
28	Bollapadu	H.P	25	88	7.6
29	Modumuru	H.P	17	56	7.7
30	Elamoru	D.W	4.0	52	7.7
31	Elamuru	H.P	17	10	7.6
32	Peddamudu	H.P	25	10	7.7
33	Pammaru	H.P	25	10	7.7
34	Oyyuru	H.P	20	10	7.7
35	Chdavaram	B.W	53	10	7.5
36	Pennamuluru	H.P	35	10	7.9
37	VenMuddurukuru	H.P	20	92	7.7
38	Krishn river	River		10	7.2
39	Mudduru	B.W	70	10	7.5
40	Vankuru	D.W	3.7	108	7.9
41	Kankipadu	H.P	30	25	7.6
42	Royyuru	H.P	15	40	7.6
43	Vallurupalem	H.P	20	10	7.6
44	Valluru	H.P	10	56	7.7
45	Kapileshwaram	H.P	12	72	7.8
46	Uranki	H.P	27	80	7.8
47	Juthovaram	H.P	25	40	7.8
48	Nidumolu	D.W	2.5	44	7.9

49	Gattapalem	D.W	1.9	25	7.5
50	Gudore	D.W	2.3	120	7.8
51	Sultanpur	D.W	2.1	56	8.2
52	Kalkenpeta	D.W	3.4	130	7.8
53	Gollapalem	D.W	2.2	63	8.4
54	Chellapalle	D.W	2.1	62	7.8
55	Kudali	H.P	25	51	7.8
56	Penumaru	H.P	8	42	7.8
57	Paddapudi	H.P	10	10	7.6
58	Movva	H.P	28	10	7.6
59	Krishnapuram	D.W	2.1	20	7.8
60	Meduru	H.P	8	24	7.5
61	Chilkapudi	D.W	2.5	122	7.9
62	Manganpudi-1	D.W	4.5	16	8.3
63	Manganpudi-2	D.W	3.6	86	7.6
64	Pedana	D.W	1.9	125	8.3
65	Nadupura	D.W	1.3	10	8.5
66	Vallamunudu	D.W	1.8	108	8.0
67	Kavataram	D.W	1.7	18	7.9
68	Gollavaluru	H.P	10	20	7.9
69	Angaluru	H.P	13	25	8.0
70	Avanigadda	H.P	8	10	7.6
71	Nayalanka	H.P	26	10	7.6
72	Vizayawada	H.P	13	10	7.7
73	Ramavarepalle	H.P	10	60	7.8
74	Teluprolu	H.P	12	53	8.0
75	Paramulu	H.P	13	10	7.4

South Krishna Delta

S.No	Name	Type	Depth	Nitrate	pH
1	Peddaravuru-1	H.P	8.0	22	7.5
2	Peddaravuru-2	H.P	9.0	25	7.8
3	Pedolguru	H.P	25	20	7.6
4	Jampani	H.P	10	42	7.8
5	Vemuru	H.P	10	10	7.6
6	Kolluru-1	H.P	17	10	7.6
7	Kolluru-2	H.P	10	10	7.6
8	Denapudi	H.P	33	24	8.2
9	Vellaturi	H.P	9	10	7.8
10	Bhattiprolu	D.W	10	25	7.8
11	Kodimaru	D.W	3.3	24	8.2
12	Chevali	D.W	4.6	28	7.8
13	Peravil	D.W	4.5	20	7.6
14	Mulpur	D.W	4.4	10	7.6
15	Kuchipidi	D.W	4.4	50	7.5
16	Ansalkuduru	H.P	9.0	35	7.5
17	Jagarlamudi	H.P	12	62	7.9
18	Garuvupalem	D.W	5.0	68	7.9
19	Vadlamudi	D.W	11	42	7.4
20	Narakoduru	D.W	9.5	33	7.8
21	Chebrolu	D.W	4.4	105	7.8
22	Mancherla	D.W	4.0	31	7.4

23	Munnupalle	H.P	8.0	62	7.5
24	Inturu	D.W	2.0	40	7.8
25	Gollapallu-1	D.W	3.8	68	7.8
26	Gollapallu-2	D.W	3.0	75	8.2
27	Panchalva	D.W	3.5	110	8.3
28	Marlaluru	D.W	3.7	40	8.1
29	Pedipudi	D.W	3.7	10	7.8
30	Nanadivelugu	D.W	4.0	10	7.8
31	Thuruluru	H.P	8.0	10	7.5
32	Chivaluru	H.P	10	30	7.8
33	Attola	H.P	13	52	7.8
34	Kolakaluru	D.W	3.0	38	7.6
35	Kazipet	HP	50	53	7.4
36	Imani	H.P	27	48	7.4
37	Vallbhapuram-1	H.P	23	30	7.8
38	Vallbhapuram-2	H.P	20	54	7.5
39	Munnad	H.P	16	60	7.5
40	Chakkavayapadam	D.W	8.4	10	7.4
41	Chadalavada	H.P	23	10	7.4
42	Anathavaram	H.P	24	10	7.4
43	Channavaranur	D.W	15	135	7.5
44	Chennasajula	D.W	20	105	7.8
45	Peddabajuluru	H.P	20	22	8.0
46	Sunduru	H.P	16	10	8.0
47	Vativeru	H.P	7	10	7.8
48	Eolapalli	H.P	8	20	7.6
49	MOlluluru	H.P	25	35	7.6
50	Mopuru	D.W	4.3	10	8.4
51	Govada	D.W	4.5	38	8.2
52	Rajavolu	D.W	3.8	32	8.0
53	Nagaram	D.W	3.3	30	8.0
54	Repallr-1	H.P	15	22	8.2
55	Repallr-2	H.P	15	10	8.0
56	Petaru	D.W	3.0	18	8.0
57	Karumuru	D.W	3.0	40	7.5
58	Pallevaru	H.P	7.0	10	7.6
59	Dholpuri	D.W	3.0	10	7.5
60	Ponnuru	H.P	7.0	10	7.8
61	Sandoavow	D.W	2.8	10	7.6
62	Repdipalem	D.W	2.9	10	7.8
63	Alluru	D.W	1.7	10	7.5
64	Nizampatnam	D.W	8.0	44	7.8
65	Achutaparam	D.W	2.9	20	8.2
66	Garuvupadam	D.W	3.0	32	7.8
67	Tenali-1	H.P	8.0	10	7.5
68	Tenali-2	H.P	8.0	10	7.4
69	Tenali-3	H.P	10	10	7.4

D.W : Dug well   B.W : Bore well   H.P : Hand pump
Depth in meters,   Nitrate concentration in mg/l pH = -logH$^+$

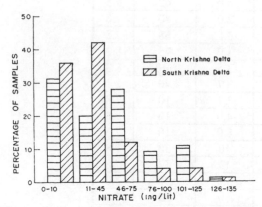

Figure 2: Sample percentage based on different ranges of nitrate concentration

Result and discusssion

The studies carriedout on the nitrate contents have shown that concentration varies from 5 to 135 mg/l. The nitrate concentration in different ranges, classified these water samples into six categories and shown in Fig.2. This figure show, 31% samples of north Krishna delta and 36% of south Krishna delta have low nitrate (<10 mg/l); 20 % samples of north and 42% of south in between 11 to 45 mg/l; 28 % samples of north and 12 % of south are in between 46 to 75 mg/l; 9 % samples in north and 4 % in south have 76 to 100 mg/l and seems to be at a high pollution level and the water can be used other than drinking purposes; 11 % in north and 4 % in south have a high value of nitrate (101-125 mg/l), and extremely high nitrate content (>126 mg/l) is reported in 1 % of groundwater samples in both the regions. The isolevels of nitrate have been drawn in 1: 250,000 scale and shown in Fig.3. This figure indicate that the behavior of nitrate enrichment is high in Gudiwada, Gannavaram and Muchlipatnam areas of north Krishna delta. The south Krishna delta which is near to sea coast are free from nitrate pollution. In general the possible sources of nitrate that lead to nitrite in groundwater are nitrogen rich sediments, interaction of groundwater with nitrogen rich industrial waste, inputs of organic nitrogen into soil, biological denitrogen fixation by micro organisms, inputs of human and animal waste, water in unused dugwells, stagnate water and nitrogenous

inorganic fertilizers etc (Burden, 1986, Hammer, 1986, Dudley, 1990, Barnes and others 1992). During the field investigation, no such industries were found which can transport the nitrate rich industrial waste in the studied areas. Krishna delta is known for its fertile soil which is very much favourable for the cultivation of mainly paddy and sugarcane. The inorganic nitrogenous fertilizers are used to increase the growth rate of these agricultural production. The common inorganic nitrogen fertilizers which are widely used in the region are urea, calcium ammonium nitrates, ammonium chloride and calcium phosphates etc. The nitrogen content in these fertilizers are high and vary from 20 to 50 %. Mostly these fertilizers are soluble in water and easily releases nitrogenous mass (Houzin etal, 1986 Burt etal 1993). In chemical process, fertilizer react with water/moisture and decomposes into amino acid. These amino acids are degraded to ammonium sulphates and ammonium amines which in tern oxidize to nitrites. In atmospheric condition nitrites are very unstable and soon converted into nitrates. Inview of the nitrates solubility in water, they rapidly increase the nitrate pollution level in the groundwater. The range of nitrate in different water sources and their mean value are shown in Tab.2. On the basis of Fig.2 and Tab.2, it is observed that the groundwater of north Krishna delta is more polluted than south. Field study and personal communication with concern people indicate that high quantity of inorganic nitrogen fertilizers was used in this region. The nitrate data also indicate pollution level is more in dugwells compared to handpumps / bore wells (Table.2). It is mainly because of poor dugwells structure. In practice nitrogen fertilizers are supplied in the agriculture field at a period before the crop grows but some time the crop yields may not be as high as anticipated because of unrealistic yield or adverse weather condition and this can result in runoff and leaching of residual nitrate in the soil and inturn pollute the groundwater.

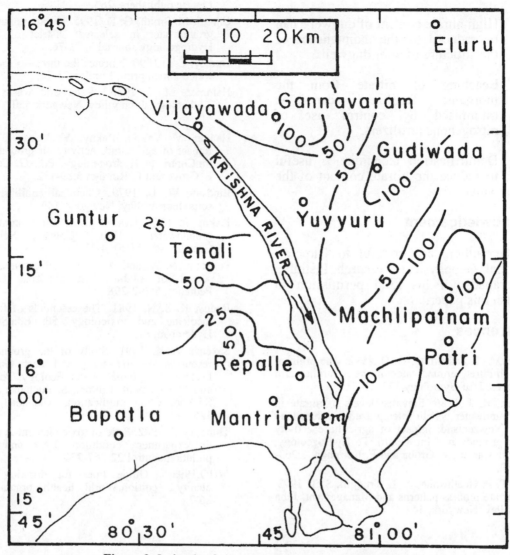

Figure 3: Isolevels of nitrate concentration in Krishna delta

Table 2: Range of Nitrate and their mean values in water samples of Krishna delta.

	Number of samples	Range of Nitrate	Mean value of Nitrate
(1) North Krishna Delta			
Dug well	31	10-130	61
Bore well	5	10-116	31
Hand pump	38	10-116	35
(2) South Krishna Delta			
Dug well	34	10-135	40
Hand Pump	35	10-62	25

Concentration of Nitrate in mg/l

175

Recommendations

1. High nitrate content of dugwells can be reduced by the maintenence of the structure of poor dugwells.

2. Leaching of nitrate from the inorganic fertilizers can be minimised by control use of nitrogenous fertilizers.

3. Denitrification technique is useful to reduce the nitrate content of the water.

Acknowledgement

The authors are grateful to Director, National Geophysical Research Institute, Hyderabad, for his kind permission to publish this paper.

References

Barnes, C. J. and Smith, G. D. 1992. The orgin of high nitrate groundwaters in the Australian arid zones. J. of Hydrology 137, 181-197.

Burden, R. J. 1986. Elevated levels of nitrate in groundwater beneth inten.grazed pausture land in Newzealand, impact of agriculture activity on groundwater. Int. Contri. To Hydrogeology, Edn. Castany.F, Groba and E. Romingn, 5:263-269.

Burt, T. P. Heathwaite, A. L. Trudgill, S. T. 1993. Nitrate process patterns and Management. John Willey, Newyork, 444.

Canter, L. W. 1997. Nitrate in groundwater. Lewish publishers, 263.

Chettri, M. Smith, G. D. 1995. Nitrate pollution in groundwater in selected District of Nepal. Hydrogeology Journal 3, 71-76.

Dudley, C. J. 1990. Nitrates the threat to food and water. Green print, London 71-76.

Hammer, M. J. 1986. Water and waste water Technology, John villey, Newyork 550.

Hauzin, V. Vavraj, Pekny, V. Veva, J. 1986. Impact of agriculture activity on groundwater. Int.Contri. to Hydrogeology. Edi. G. Castany, E. Grova and E. Romign 5:89-132.

Lindsay, W. L. 1979. Chemical equilibrium in soils. Jhon Willey, Newyork, 149.

Piskin, R. 1973. Evaluation of nitrate content of groundwater Hall Couty Nebraska. Groundwater 11(6): 4-13.

Perlstein, N. A. and Attala, R. 1976. Neurologic sequel of plubism in children. Clinical Pediatics, 5:292-298.

Reddy, K. S. N. 1981. The essentials of forensic medicine and toxicology. 5th Edn.Sugune, Hyderabad, 81.

Saxena, V. K. 1991. Study of the groundwater resources of the west and east Godavari District, A.P, India. Water Ecology, Pollution and Management, Editors, S. Pitchaiah and S. S. Rao, Chug publication, Allahabad, India 125-147.

Thind, G. S. 1982. Role of toxic element in human on experiment condition. J.Air and water pollution control 22:267-270.

WHO,1993 . Guide lines for drinking water quality,2 edition.World health organization, 267.

Intl. Conf. on Sustainable Development and Management of Groundwater Resources
in Semi-Arid Region with Special Reference to Hard Rock, (IGC 2002),
M. Thangarajan, S.N. Rai & V.S. Singh (Eds.)

Towards quantification of ion exchange in a sandstone aquifer

John Tellam, Harriet Carlyle[1], Hamdi El-Ghonemy[2],
Karen Parker, and Richard Mitchener[3]

Earth Sciences, University of Birmingham, Birmingham, B15 2TT, UK, email: J.H.Tellam@bham.ac.uk
[1] Now at Earth Tech Engineering Limited, Barnsley, S75 3DL, UK
[2] Now at JacobsGIBB, Stockport, SK3 0XF, UK
[3] Now at SLR Consulting, Alvechurch, Worcestershire B48 7DD, UK

Abstract

Ion exchange is a very important process in determining the migration of many solutes through aquifer systems. However, it is rare that attempts are made to quantify its role. This paper outlines recent work on Na/K/Ca/Mg cation exchange reactions in the English Triassic Sandstone, a fluviatile/aeolian red-bed sequence whose main exchange phases are clay minerals and Mn and Fe oxyhydroxides.

Initial studies concentrated on the macroscopic behaviour during sea water intrusion in NW England. Standard and new laboratory measurements on core samples indicated that the cation exchange capacity is ~ 1 meq/100g, and that the Gaines-Thomas (GT) convention provides an adequate but not perfect description of exchange over a limited range of sorber compositions. Using these results and a reactive transport model, it proved possible to successfully predict the breakthrough curves at a coastal site over a period of around 40 years for which field data exist.

A second study investigated the exchange behaviour of the sandstones in detail. Major cation exchange was found to be much better-characterized by a power function than by the standard convention expressions. A constant selectivity coefficient could be derived for the whole range of sorbed site compositions investigated (~10-90%), and from the results it was possible to derive inferred sorbed phase activity coefficients.

A third study has started to investigate the exchange properties of the components of the sandstone in an attempt to understand the origin of the power function and to develop a means of predicting exchange properties from mineralogical data. The exchange behaviour of 4 clay minerals (kaolinite, smectite, illite, and chlorite) have been examined, and in each case the power function provided an excellent description. Mixtures of the minerals give rise to properties which vary in non-linear but smooth fashions with mixing fraction. It is thus possible to predict empirically the ion exchange properties of a rock containing a mixture of clay minerals. It is hoped to extend this work to MnO_2.

Preliminary modelling work has also been completed on up-scaling from laboratory to field scales. This shows that in general perfect GT exchangers in parallel cannot be replaced by a single GT exchanger. It appears that in practice one GT exchanger could be used where the properties of the individual exchangers are similar: when they are not, a power function can be used, but even the latter fails when the exchangers have very different properties.

A final study has determined that Mn oxyhydroxides provide a significant proportion of the cation exchange capacity in some of the sandstones; provided data on cation exchange capacity distributions; and has determined that the exchange capacity of the sandstones is only weakly correlated with permeability and porosity.

Key Words: Ion exchange; groundwater; aquifer; Triassic Sandstone

Introduction

Ion exchange is a very important process in determining the migration of a number of solutes, including pollutants, through aquifer systems. It affects the relative proportions of ions, and by doing so can induce other reactions to occur resulting in the loss or gain of solute to the groundwater and, often, a change in pH: for example, displacement of fresh, calcite-saturated groundwater by a Na-rich solution (e.g. landfill leachate or sea water) will usually result in the increase of Ca in solution and hence the precipitation of calcite and a fall in pH. Such changes in chemistry will often have implications for migration of pollutants. Ion exchange is also directly involved with attenuating a number of pollutant species (e.g. NH_4^+).

Despite the importance of the process, few attempts have been made at quantification of ion exchange in the hydrogeological context: the studies associated with the Borden site in Canada (e.g. Dance and Reardon (1983) and Reardon et al. (1983)) and those undertaken by Appelo and coworkers (e.g. Appelo (1994); van Breukelen et al. (1998)) are exceptions, and a great deal more work has been undertaken on soils (e.g. see Sposito (1981); McBride (1994)).

This paper outlines recent studies aimed at quantifying the ion exchange behaviour of the English Triassic Sandstones, a fluviatile/aeolian red-bed sequence whose main exchange phases are clay minerals and Mn and Fe oxyhydroxides.

The first study described undertook laboratory determination of ion exchange parameters for sandstone samples from northwest England (Carlyle, 1991). The data were interpreted using a standard soil science approach, and then used in reactive transport computer modelling studies of the intrusion of estuary water into the aquifer over a 40 year period. Although surprisingly successful in reproducing the breakthroughs of the various cations at the pumping wells, the study showed that although the standard ion exchange model proved satisfactory, it provided a far from perfect description of the exchange processes as observed in the laboratory. As a result, a second study examined the exchange behaviour of sandstone samples in detail (El-Ghonemy,

1997), and this resulted in a new empirical description of exchange for the sandstones. A third study is attempting to determine how the various exchange phases contribute to the overall exchange behaviour of the rock, and hence whether knowledge of the mineralogy of the sandstones can enable exchange properties to be predicted (Parker et al., 2000). The final study described here has attempted to map the correlations between hydraulic and geochemical properties of the sandstones, and this work has included examining cation exchange capacities (Mitchener, PhD thesis in prep.). In all cases, only major cation (Ca, Mg, Na, K) ion exchange has been considered.

The English Triassic Sandstone

In the areas studied, the Triassic sandstone typically contains (e.g. Gillespie, 1987): quartz clasts - 50-70 percent of whole rock volume; feldspar clasts (mainly K-feldspar) - 3-6%; calcite clasts and cement - 1-3%; and clay minerals (illite, chlorite, kaolite, smectite, mixed layers) - < 5%. Iron oxyhydroxides coat almost all the grains, with Fe comprising 1-2% of the rock, and Mn oxyhydroxides are also common. Pyrite and various accessory minerals are also occasionally present (Travis and Greenwood, 1911), though pyrite was not observed in any of the samples used in the studies.

The hydraulic conductivity as measured in the laboratory typically averages ~ 1m/d, but ranges from < 10^{-3} to > 10 m/d. Porosity averages around 25%, and specific yield is probably ~ 10-15%, though good data are few. The sandstone is fractured, but often the regional hydraulic conductivity is not much greater than the intergranular conductivity.

The Application of the Gaines-Thomas Exchange Model in Predicting Groundwater Chemistry in a Zone of Estuary Intrusion

This study attempted to apply an ion exchange model widely used in soil science in a region of northwest England where estuary water has been intruding the sandstone aquifer since around 1930. The aquifer underlies the town of Widnes. Abstraction from industrial boreholes in the town and industrial and public supply boreholes to the north have resulted in a northward-directed head gradient and the intrusion of brackish water (Cl < 10,000 mg/l) from the Mersey Estuary (Howard, 1988; Tellam and Lloyd, 1986).

Determination of Exchange Parameters

Sixteen cylindrical plugs, representing the main lithologies present, were obtained from a 239m borehole core in Widnes. The plugs were analysed intact using the method of Reardon et al. (1983): vacuum saturation with deionized water, removal of porewater using centrifugation, analysis to determine porewater concentrations in equilibrium with the exchangers, saturation with 1M LiCl, centrifugation, and analysis to determine cation exchange capacity (CEC) and the composition of the exchange sites. From the porewater concentrations and the compositions of the exchange sites, selectivity coefficients were obtained assuming the Gaines-Thomas (GT) convention (Gaines and Thomas, 1953):

$$K_{A2+/B+} = \frac{(B^+)^2 A^{2+} X}{(A^{2+})B^+ X^2}$$

where ( ) represents activities calculated using the extended Debye-Huckel equation, and $A^{2+}X$ is the equivalent fraction of ion A^{2+} on the exchanger (sorbed A^{2+} /CEC, both in equivalents).

Two "runs" of experiments were completed (Table 1): between the runs the cores were saturated with a saline solution, and hence the exchange conditions are different in Run 1 and Run 2 as indicated in Table 2. Hence the experiments cover a very wide range of conditions. No pH was control was attempted in the experiments, but was always in the range 6-6.5. Field sample pHs for 1979-1980 were in the range 6.0-8.0, with most values lying between 6.7 and 7.4.

A flushing method was also used to determine ion exchange parameters (cf. Kool et al., 1989). This involved passing a set of solutions through the core plugs and recording the breakthrough curves. From these data the mass transfers between solution and exchangers can be calculated using a direct search method as the resulting equation is a fifth-order polynomial. CEC estimates obtained using the 1 M LiCl method described above were then used with the mass transfer data to calculate selectivity coefficients assuming a particular ion exchange model (in this case GT). This approach has the advantages that less drying and porewater extraction is needed, and the validity of the chosen ion exchange model is directly determined.

Results

The results from the standard method are shown in Table 1, and have similar ranges to those given by Appelo and Postma (1993). The flushing method produced similar parameters on average. Close examination of the data sets indicated that the ion exchange parameters vary with exchange site composition. In the case of the flushing method it was sometimes necessary to make changes (mean 13%, minimum 3%, maximum 37%) to the cation exchange capacity from one flush to another in order to obtain an interpretation. This problem suggests that the GT model, although often producing a satisfactory description, is not perfect, and this observation lead to the further investigations described in Section 4.

Modelling of the Field System

Pumped water Cl and hardness data spanning a period of around 40 years were available for a number of boreholes in Widnes, together with full major ion analyses for 1980/1. Using the exchange parameters obtained from the laboratory experiments, a one-dimensional reactive transport mixing cell model was used to predict breakthrough at the field scale. The breakthrough patterns thus obtained, assuming GT/constant CEC exchange with calcite equilibrium, significantly overestimated the concentrations of the divalent ions and the 1980 sulphate concentration, and underestimated the 1980 alkalinity concentration. Inclusion of a representation of sulphate reduction in the Estuarine alluvium failed to reproduce the 1980 HCO_3 and pH values. However, ^{34}S and ^{18}O evidence from a previous study (Barker et al., 1998), indicates that partial CO_2 degassing occurs from the alluvium following sulphate reduction. By including these mechanisms, and using GT exchange with averaged laboratory parameters, good matches for SO_4, HCO_3, pH, and cation concentrations were obtained. It was concluded that the GT/constant CEC model with averaged laboratory parameter values can be used successfully when predicting estuary water intrusion in this

Table 1: CEC and K values for standard method experiments (sample numbering scheme: first 3 alphanumeric characters indicate depth; A, B signify different samples from the same depth)

Sample	CEC (meq l^{-1})	Run	$K_{Ca/Mg}$	$K_{Ca/Na}$	$K_{Ca/K}$	$K_{K/Na}$	CaX_2	NaX
10a	2.30	1	5.63	0.22	0.10	1.51	0.73	0.02
10b	1.44	1	0.87	0.08	0.22	0.60	0.49	0.01
10cA	1.83	1	0.95	0.43	23.68	0.13	0.58	0.03
10cB	1.79	1	0.77	0.74	1.17	0.79	0.59	0.02
20a	1.51	1	1.63	0.03	0.02	1.25	0.50	0.03
20bA	1.15	1	0.87	0.16	0.04	2.05	0.57	0.03
20bB	1.71	1	0.43	0.94	0.03	5.79	0.59	0.02
35b	1.21	1	1.56	7.34	0.16	6.82	0.46	0.01
40aA	0.85	1	1.03	0.53	0.06	3.10	0.59	0.02
40aB	0.97	1	0.96	0.28	0.06	2.18	0.60	0.01
40b	1.00	1	0.84	0.32	0.11	1.73	0.57	0.03
55aA	0.98	1	0.83	0.59	0.04	4.04	0.62	0.03
55aB	0.84	1	1.10	0.68	0.11	2.51	0.61	0.02
59a	0.95	1	1.01	0.60	0.11	2.34	0.59	0.02
78a	0.92	1	2.84	0.03	0.04	0.90	0.53	0.04
78b	0.09	1	0.47	0.00	0.00	1.56	0.29	0.23
10cA	2.20	2	0.66	0.45	0.09	2.23	0.37	0.40
10cB	1.62	2	0.88	0.76	0.05	4.11	0.33	0.47
20	1.17	2	0.89	0.67	0.19	1.87	0.34	0.42
20bB	2.00	2	0.68	0.59	0.14	2.07	0.37	0.42
35b	0.69	2	0.77	0.31	1.00	1.79	0.27	0.45
40aA	0.71	2	0.39	1.17	0.05	4.70	0.23	0.42
40aB	0.83	2	0.38	0.64	0.14	2.11	0.24	0.44
40b	1.44	2	2.10	30.41	0.39	8.85	0.79	0.12
55aA	0.60	2	1.27	75.7	0.26	17.24	0.53	0.11
55aB	0.92	2	1.02	12.76	0.17	8.61	0.57	0.17
59aB	0.69	2	3.67	476	0.51	30.6	0.80	0.13
78	0.81	2	1.29	0.88	0.17	2.31	0.35	0.44
78b	0.81	2	1.37	1.24	0.23	2.33	0.33	0.45
Mean*	1.20		0.92	0.51	0.20	2.90		
St. Dev.*	0.46		0.33	0.35	0.27	2.26		
n	19		29	26	28	27		
Geom mean *	1.03		0.85	0.39	0.12	2.32		
Total Range*	0.6-2.3		0.38-5.6	0.0-1.2 (7.3)	0.0-1.2	0.13-8.9		
AP(1993) Range@			0.4-4.0	2.8-11.1	0.06-0.7	4.0-6.7		

* Ignoring outliers indicated by italics. Zeroes ignored for geometric mean calculations.
@ Range indicated by Appelo and Postma (1993).

Table 2: Porewater concentration ranges in experiment runs 1 and 2

	Ca	Mg	Na	K
Run 1	6-24 mg/l	3-11 mg/l	5-65 mg/l	4-12 mg/l
Run 2	22-533 mg/l	6-213 mg/l	313-2058 mg/l	13-70 mg/l

aquifer at a regional scale and over long time scales, despite the numerous assumptions necessary. This result agrees with the conclusion of Appelo (1994) who also studied a large-scale flow system.

Investigation of the Exchange Properties of Sandstone Samples in the Laboratory

The previous investigation had suggested that the GT model, although appearing to perform satisfactorily when applied in a regional investigation, clearly did not describe the exchange processes fully. Hence a second study was undertaken to examine the exchange behaviour of the sandstones in more detail. Four intact core plug samples were analysed to determine how their exchange properties varied as a function of exchange site composition. A search was then made to obtain an empirical model which would provide a better description for the ion exchange properties of the samples.

Methods

The samples were from core from the same aquifer as the previous study. The basic approach followed that of Jensen and Babcock (1973). The plugs were flushed with a solution of known cation ratio until equilibrium was attained: solutions containing only two cations were used. The plug was then flushed to equilibrium with a concentrated solution in order to displace the previously sorbed cations. From the data collected, the selectivity coefficients were calculated for a wide range of exchange site compositions.

Results and Interpretation

Selectivity coefficients were found often to vary considerably as a function of exchange site composition. Figure 1 shows example results for GT selectivity coefficients.

One way to interpret the data is to assume that the variation in selectivity coefficients is due to variation in sorbed ion activity coefficients. If this is done, then the variation in sorbed ion activity coefficient is of the form shown in Figure 2. Similar activity coefficient variation was found for most of the data sets, and in principle this approach might be used with a constant selectivity coefficient.

However, it was found that a simple empirical power function fitted the data very well:

$$K_{A2+/B+} = \frac{(B^+)^2 \left[A^{2+} X \right]^n}{(A^{2+}) \left[B^+ X^2 \right]^n}$$

Figure 3 shows some example data: a power function relationship is clearly indicated. Power functions have been used before to describe ion exchange systems, notably by Rothmund and Kornfeld (1918;1919), Walton (1949), Langmuir (1981; 1997), and Bond (1995), but have rarely been used in the hydrogeological context. The thermodynamic justification is limited (e.g. see Langmuir (1997)), but power functions can derived by assuming that the sorbed cations are effectively represented by an ideal solid-solution (Garrels and Christ, 1965). They also appear to be appropriate in heterogeneous exchange systems, a topic investigated in the study described in Section 5.

The Exchange Behaviour of Individual Exchange Phases in the Sandstone

To investigate the reasons for the exchange behaviour and to attempt to see if exchange properties could be predicted given a knowledge of the mineralogy of the rock, a study is presently examining the exchange properties of some of the main exchange phases in the sandstone. The study has also undertaken a preliminary modelling investigation of issues associated with the up-scaling of results from the laboratory scale to the field scale.

Laboratory Methods

The exchange phases in the sandstone are clays and oxyhydroxides. To date, the clay phases only have been considered. Because it is not feasible to extract enough pure clay from the sandstones, samples of kaolinite, chlorite, montmorillonite, and illite were purchased from the Clay Minerals Repository, Missouri, USA. The exchange properties of the clays were determined again using the Jensen and Babcock (1973) method, but this time in batch reactors. Experiments involved examining single clays, and binary and ternary mixtures of the clays. XRD indicated that the clay samples contained no recognizable amounts of

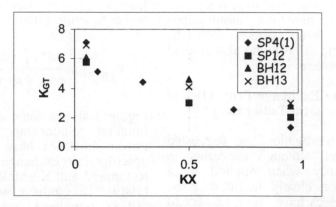

Figure 1: The variation of the GT selectivity coefficient (K_{GT}) as a function of the equivalent fraction of K^+ on the exchange sites (KX) of four Triassic sandstone samples undergoing K/Na exchange.

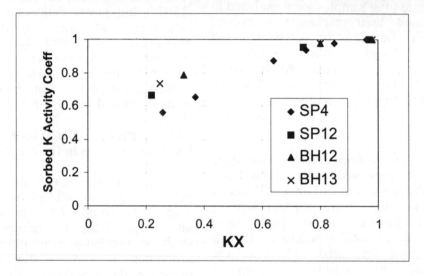

Figure 2: The variation of sorbed phase activity coefficients as a function of the equivalent fraction of K^+ on the exchange sites (KX) of four Triassic sandstone samples undergoing K/Na exchange. Values were calculated using the method of Argersinger et al. (1950).

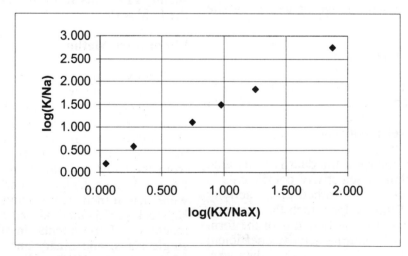

Figure 3: The power function relationship for K/Na exchange on Triassic sandstone sample SP4

other minerals. Again, pH was not adjusted, but was measured at all stages.

Laboratory Investigation Results

In all cases the GT, Gapon, and Vanselow selectivity coefficients varied as a function of the exchange site composition (Figure 4). However, the exchange on all the pure and all the mixed clay samples was well described by a power function. Figure 5 shows example results. The power function parameters vary smoothly as a function of mixture proportions, as shown in Figure 6. It would therefore appear that there is a possibility that the exchange properties of rocks containing only clay exchange phases may be estimated using data on the proportions of the clay components. The study continues.

Preliminary Up-Scaling Investigation

To investigate the issues associated with the up-scaling of laboratory results to the field scale, modelling work using PHREEQM (Appelo and Postma, 1993) has been undertaken. The basic approach is to model the breakthrough from two stream tubes with different ion exchange properties, to combine the effluent stream, and then to interpret the mixed effluent. The combined effluent water is taken to represent the water obtained from, for example, a sampling well. This study has shown that for a series of perfect G-T exchangers:

(i) the chemical changes observed in the "sampling well" can be reproduced adequately using a GT description with averaged exchange capacities and selectivity coefficients but only if the exchange properties of each stream tube are very similar;

(ii) for stream tubes where the exchange properties differ significantly from each other, the GT description no longer is satisfactory, whatever the exchange properties assumed;

(iii) a power function description will describe the breakthrough curves very well for cases where the geochemical properties of the stream tubes are very different;

(iv) the power function description breaks down where the difference in the exchange properties of the stream tubes is extremely large (usually unrealistically large) - i.e. the power function description appears to be only approximate for heterogeneous systems.

Again, the study continues.

The Role of Oxyhydroxides in Exchange in the Sandstones, and Correlations Between Hydraulic and Geochemical Properties

If clays are an important exchange phase in the sandstones, it is possible that there is a negative correlation between permeability and cation exchange capacity. Accordingly an investigation is being undertaken.

Methods

Around forty samples from the core of a borehole on the Birmingham University campus have been gently broken up and analyzed to obtain cation exchange capacity. Core samples from immediately adjacent to the geochemical samples were analyzed to determine porosity and hydraulic conductivity using water saturation and falling head methods. In addition, gently broken up samples were leached with hydroxyl ammonium chloride in 25% ethanoic acid solutions in order to remove oxyhydroxides. The solutions were analyzed for cations, and the cation exchange capacity was determined before and after leaching (Mitchener, PhD thesis in prep.).

Results

Cation exchange capacity varies irregularly as a function of depth: values range from 1 to 20 meq/100g. CEC correlations with gamma emission and clay type were surprisingly weak, but CEC/porosity correlation was significant at the 99% confidence level (Spearman rank correlation; negative correlation). As expected, no correlation was found between CEC and fraction of organic carbon, as the sandstones contain very little organic matter ($< 0.1\%$). The correlation between cation

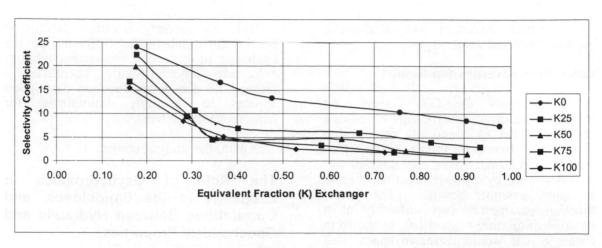

Figure 4.: The variation of GT selectivity coefficients as a function of equivalent fraction of K^+ for various mixtures of kaolinite and illite (K25 = 25% kaolinite).

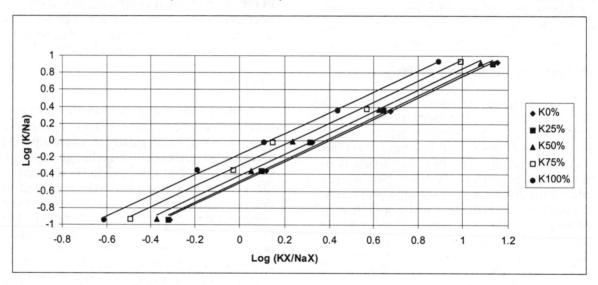

Figure 5: The power function relationships for K/Na exchange for various mixtures of kaolinite and illite (K25 = 25% kaolinite).

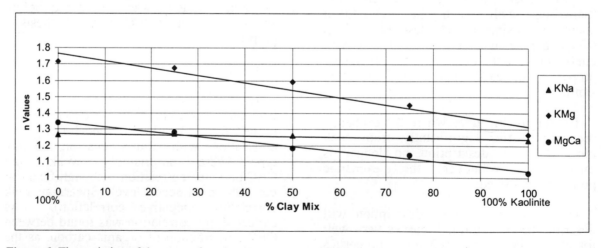

Figure 6: The variation of the power function exponent as a function of mix % for kaolinite/illite mixtures for three binary exchange systems - K/Na, K/Mg, and Mg/Ca.

184

exchange capacity and hydraulic conductivity is weak, though a Spearman rank correlation test suggests that it is still significant at the 99% level for the sandstone with a medium saturation colour, though not for lower or higher colour saturations. The weakness of the hydraulic conductivity correlation appears to arise at least partially because the clays are not located in the pore throats of the sandstone, but are found more commonly lining the pores as evidenced by electron microscope examination. The leaching experiments indicate a strong correlation between change in CEC following leaching and the amount of Mn removed by leaching, but only in the darkest coloured sandstones. For at least the darker coloured sandstones, this implies: (i) that it will be necessary to investigate the exchange properties of the oxyhydroxides if the sandstone system is to be fully understood; (ii) that, because the power function describes the rock and the clay exchange behaviours well, it will probably also describe the oxyhydroxide behaviour too; and (iii) that the overall behaviour of a significant part of the exchange phases will be pH dependent. The importance of the Mn phase also suggests that the attenuation capacity of the sandstone to a plume of reducing solutes (e.g. landfill leachate) will be significantly decreased as reductive dissolution occurs and part of the exchange capacity of the sandstone is destroyed. However, the overall importance of the Mn oxyhydroxides does vary. being more significant in the darker coloured sandstones.

Conclusion

At the laboratory sample scale, none of the usual models of exchange behaviour provide a really good description of ion exchange in the Triassic Sandstones of England. However, a power function description is significantly better. Of the main exchange phases - clays and Mn oxyhydroxides - the clay exchange behaviour can also be well described by a power function: although the work has not yet been done, it would seem likely that the Mn phase behaviour will also be found to be reasonably described using a power function (though there are good mechanistic models for sorption in oxides (e.g. Dzombak and Morel, 1990)). There is only a weak correlation between hydraulic conductivity and cation exchange capacity in the sandstones.

At the field scale it proved possible to model the breakthrough of estuary water over a 40 year period using the laboratory data and a Gaines-Thomas convention description. However, this may well have been fortuitous, especially given the very crude hydraulic description used in the model, and it is suspected that a power function model would have produced at least as good a result. Preliminary model experiments seem to suggest that a power function description may arise in cases where several exchangers with rather different properties are present, and this is consistent with the laboratory findings that power function descriptions are valid for clay mixtures, and for the sandstones themselves, which are clearly heterogeneous exchangers.

References

Appelo, C.A.J., 1994. Cation and proton exchange, pH variations, and carbonate reactions in a freshening aquifer. Water Resources Research 30, 2793-2805.

Appelo, C.A.J., and Postma, D., 1993. Geochemistry, groundwater and pollution. Balkema, Rotterdam.

Argersinger, W.J., Davidson, A.W., and Bonner, O.D., 1950. Thermodynamics of ion exchange phenomena. Trans. Kans. Acad. Sci., 53, 404-410.

Barker, A., Newton, R., Bottrell, S. H., and Tellam, J. H., 1998. Processes affecting groundwater chemistry in a zone of saline intrusion into an urban aquifer. Applied Geochem., 6, 735-750.

Bond, W.J., 1995. On the Rothmund-Kornfeld description of cation exchange. Soil Sci. Soc. Am. J., 59, 436-443.

Carlyle, H.F., 1991. The hydrochemical recognition of ion exchange during seawater intrusion at Widnes, Merseyside, UK. Unpublished PhD thesis, Earth Sciences, University of Birmingham. 275pp.

Dance, J.T. and Reardon, E.J., 1983. Migration of contaminants in groundwater at a landfill: a case study, 5. Cation migration in the dispersion test. J Hydrology, 63, 109-130.

Dzombak, D.A. and Morel, F.M.M, 1990. Surface complexation modelling: Hydrous ferric oxide. John Wiley & Sons, New York, 393.

El-Ghonemy, H.M.R., 1997. Laboratory experiments for quantifying and describing cation exchange in UK Triassic Sandstones. Unpublished PhD Thesis, University of Birmingham, UK, 239pp.

Gillespie, K.W., 1987. The sedimentology and diagenetic history of the sandstones in the Merseyside Permo-Triassic aquifer. Unpublished MSc Thesis, University of Birmingham, pp. 108.

Gaines, G.L., and Thomas, H.C., 1953. Adsorption studies on clay minerals, II. A formulation of the thermodynamics of exchange adsorption. Journal of Chemical Physics, 21, 714-718.

Garrels, R.M. and Christ, C.L. , 1965. Solutions, minerals, and equilibria. Freeman, Cooper & Company, California, USA, 450pp.

Howard, K.W.F., 1988. Beneficial aspects of sea-water intrusion. Ground Water, 25, 398-406.

Jensen, H.E. and and Babcock, K.L, 1973. Cation exchange equilibria on a Yolo loam. Hilgardia, 41, 475-487.

Kool, J.B., Parker, J.CV., Zelazny, L.W., 1989. On the stimation of cation exchange parameters ftrom column displacement experiments. Soil Sci. Soc. Am. J., 53, 1347-1355.

Langmuir, D., 1981. The power exchange function: A general model for metal adsorption onto geological materials. I: Tewari, D.H., Adsorption from aqueous solutions. Plenum Press, New York, 1-17.

Langmuir, D., 1997. Aqueous environmental geochemistry. Prentice Hall, New Jersey, USA, 600pp.

McBride, M.B., 1984. Environmental chemistry of soils. Oxford University Press.

Parker, K.E., Tellam, J.H., and Cliff, M.I., 2000. Predicting ion exchange parameters for Triassic sandstone aquifers. In: Sililo, O.T.N. et al. (eds.), Groundwater: Past achievements and future challenges, Balkema, Rotterdam, 581-586.

Reardon, E.J., Dance J.T. and Lolcama, J.L., 1983. Field determination of cation exchange properties for calcareous sand. Ground Water 21, 421-428.

Rothmund, V. and Kornfeld, G., 1918. Der Basenaustausch im Permutit. I.Z. Anarg. Allg. Chem., 108, 215-225.

Rothmund, V. and Kornfeld, G., 1919. Der Basenaustausch im Permutit. I.Z. Anarg. Allg. Chem., 103, 129-163.

Sposito, G., 1981. The thermodynamics of soil solutions. Oxford Clarendon Press, Oxford, 223pp.

Tellam J. H., and Lloyd, J. W., 1986. Problems in the recognition of seawater intrusion by chemical means: an example of apparent chemical equivalence. Quarterly Journal of Engineering Geology 19, 389-398.

Travis, C.B., and Greenwood, H.W., 1911. The mineralogical and chemical constitution of the Triassic rocks of Wirral. Proc Liverpool Geological Society, 11, 116-139.

van Breukelen, B.M., Appelo, C.A.J., Olsthoorn, T.N., 1998. Hydrogeochemical transport modelling of 24 years of Rhine water infiltration in the dunes of the Amsterdam Water Supply. J. Hydrology, 209, 281-296.

Walton, H.F., 1949. Ion exchange equilibria. In: Nachod, F.C. (ed.), Ion exchange theory and practice, Academic Press, New York.

Intl. Conf. on Sustainable Development and Management of Groundwater Resources
in Semi-Arid Region with Special Reference to Hard Rock, (IGC 2002),
M. Thangarajan, S.N. Rai & V.S. Singh (Eds.)

Hybridisation of qualitative and quantitative resistivity data interpretation - a technique to resolve the fresh and brackish water in aquifers in regional affected terrains (a case)

G.K. Hodlur, S.D. Deshmukh, T. V. Rao, K. P. Panthulu and R. Dhakate

National Geophysical Research Institute, Hyderabad – 500007

Abstract

Vertical electrical soundings (VES) conducted near four fresh water bore wells, two brackish water bore wells and few virgin sites were interpreted using the curve matching technique by making use of standard master curves. The interpretation which arrived at certain subsurface resistivity models were used to delineate fresh water aquifers in the quality affected villages of Navalgund Taluk of Dharwar district in Karnataka State. Apparent resistivity values measured in the field while carry out the investigation were critically analysed and a clear, easily understandable contrast in the behaviour of apparent resistivities (field values) for brackish water and fresh water aquifers region is illustrated. Comparative study of the interpreted layer parameters and the qualitative analysis was carried out and the advantage of such comparison in reducing the constraint of uncertainties while resolving the layer parameters by curve matching technique is demonstrated.

Keywords: VES, fresh and brackish water, apparent resistivity.

Introduction

National Geophysical Research Institute carried out geophysical investigations in Navalgund taluq of Karnataka State, India (figure 1), to locate sites for exploration of potable groundwater aquifers. The main objective of the investigation was to identify fresh groundwater aquifers in the region affected by the presence of brackish groundwater aquifer. A map showing the locations where geophysical investigations were carried out is shown in Figure 2. The paper projects the problems encountered with quantitative resistivity data interpretation, particularly while delineating fresh water aquifers in the brackish terrain. It presents an easily identifiable behaviour of the field apparent resistivity characters and patterns for fresh and brackish groundwater aquifers. The quantitative data interpretation has been compared with the qualitative behaviour of the field data while taking decisions on recommendations for the exploration of potable groundwater. The above study illustrates and demonstrates the advantage of blending qualitative and quantitative interpretation to over come the stigma of uncertainty cropping up with the quantitative data interpretation.

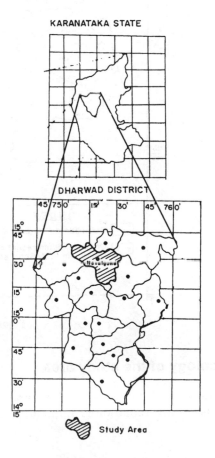

Figure 1: Study area

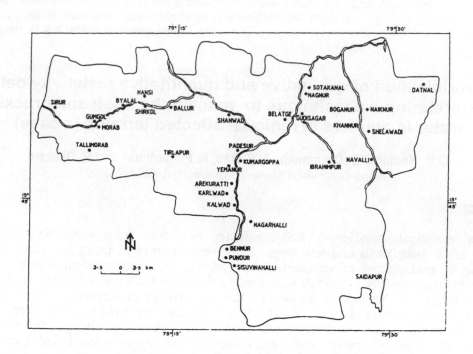

Figure 2: Drainage map of NavalGund Taluk, Dharwar district

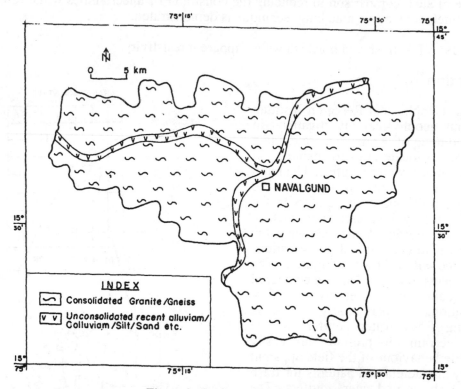

Figure 3: Geology of study area

Geology of the study area

Navalgund taluq is a flat topography without any remarkable undulation in the terrain. The taluq is drained by the stream Bennehalla and Tuprihalla, the tributaries of Malaprabha river. The taluq is covered by two formations, the archaeans and the

188

Kaladgis (equivalents of Cuddappa). The archaean forms the main exposures comprising the granites and gneisses. The Kaladgis are quartzites which occur as thin cover above the granites. The archaeans exhibit two sets of fractures in north east direction and three sets of fractures in north west direction. The major geological formations and drainage are shown in figure 3.

mtrs below ground level in dug wells and 30 to 60 mtrs in bore wells. The yield of bore wells ranges from 3000 gph to 4000 gph. The quality of groundwater in the taluq is highly brackish. Occurrence of fresh groundwater aquifers is a rare phenomena. Electrical conductivity of groundwater samples measured from various sources ranges from 3000 to 5000 micromohs. However, in some limited areas occurrence of fresh water is observed.

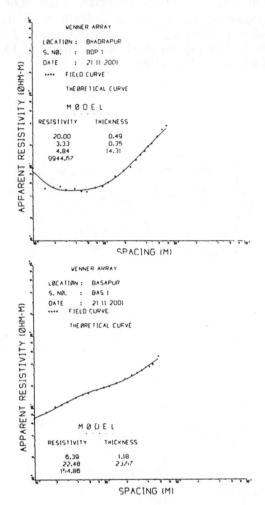

Figure 4: Resistivity sounding interpretation at Bhadrapur and Basapur(1) respectively

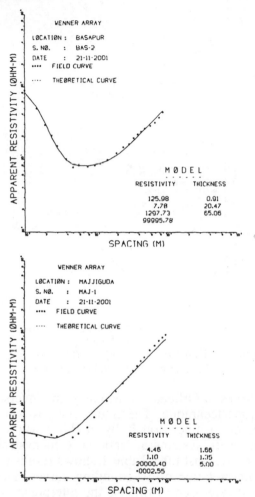

Figure 5: Resistivity sounding interpretation at Basapur(2) and Majjigudda respectively

Groundwater occurs in the unconfined condition in the weathered zone. The weathered and semi weathered zones above the archaeans are the main acquifers in the study area. In the deeper zones groundwater occurs in the fractured granites under semi-confined conditions. Thickness of the weathered zone varies from 15 to 20 metres in the study area. Depth to water table varies from 5 to 10

Geophysical investigations

Seventy four locations in Navalgund taluq were geophysically investigated for locating fresh groundwater aquifers. Few soundings were carried out in Schlumberger configuration and many others in Wenner configuration. The

189

sounding curves were interpreted by using standard mastered

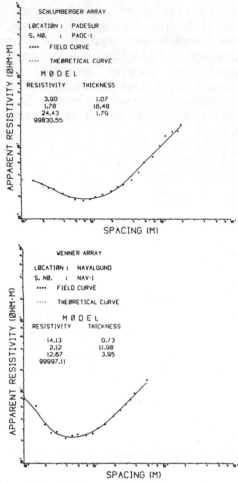

Figure 6: Resistivity sounding interpretation at Padesur and Navalguda respectively

curves of Orllena and Mooney and inverse slope technique. The interpretations were given a finer touch by making use of software systems at National Geophysical Resarch Institute. Table-1 shows results of interpretation of 10 sounding curves. The software used to finalise the interpretation was Generous Inversion Technique (Vozoff and Jupp, 1975; Verma and Panthulu, 1990).

To have an insight about the resistivities of fresh water aquifers and brackish water aquifers in the study area, six soundings were carried out; four near the fresh water borewells and two near the brackish water borewells. The results of interpretation of these soundings were taken as criteria for interpretation of the layer parameters of

the soundings carried out in the area. Figures 3, 4, 5, 6 and 7 show the sounding curves with the interpretated layer parameters of the soundings taken into consideration. The quantitative data show i.e., layer parameters reflect a double storied aquifer in the study area. A shallow aquifer occurring at a depth of 5 to 15 m possessing a resistivity range of 15 to 25 Ohm-m in case of fresh water and 5 to 15 Ohm-m in case of brackish water. The deeper aquifer occurs at a depth of 15 to 45 m possessing the resistivity range of 25 to 55 Ohm-m for fresh water and 1 to 15 Ohm-m for brackish water.

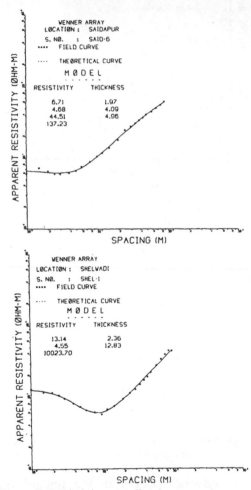

Figure 7: Resistivity sounding interpretation at Saidapur and Shelwadi respectively

Complications in resistivity data interpretation

The interpreted results of the soundings at fresh water bore well sites namely indicate Basapur, Majjiguda and

Bhadrapur 3 layered models and are shown in table-2. Sounding curves are shown in Figure 4 and 5. The resistivity of the second layer at Basapur (1) is 22 Ohm-m. Hence it should be a fresh water bearing zone and so it is. In case of Basapur (2), Majjiguda and Bhadrapur the resistivities of second layer are 7, 1.3, and 3.4 Ohm mtrs respectively. The third layered in Bhadrapur has a resistivity of 4.8 Ohm-m. and possessing a thickness of 14 m. They are followed by high resistivity layers which is the basement. In Basapur and Bhadrapur these layers are thick 20 m and 14 m, respectively. Hence these layers indicate the presence of brackish groundwater. However, they are all fresh water bore wells.

In case of Padesur and Navalgund, the resistivity models are of three and four layeres. The interpreted layer parameters are shown in table-3. Sounding curves are shown in figure 6. At Padesur the resistivity of the second layer is 1.78 Ohm-m and it is 16 m thick. The resistivity of the layer bellow this is 25 Ohm-m and it is 1.8 m thick.. Below this is the high resistivity zone. In Padesur, the resistivity as low as 1.79 Ohm-m and layer is of 16 m thickness, the presence of brackish water. The drilling also indicates a brackish water aquifer. In case of Navalgund the second layer is having 2 Ohm-m of resistivity and 12 m thickness followed by a layer of 12 Ohm-m resistivity. Hence the second layer is brackish water zone. Very low resistivity indicate the presence of brackish water in Navalgund and Padesur, but the same type of lows yield fresh water in Basapur, Majjiguda and Bhadrapur. Hence there is an uncertainty because of overlapping resistivities. The above illustrations gives a clear understanding of the constraints and difficulties which one faces while making reccommendations due to the overlapping nature of the resistivities of fresh and brackish water.

In case of Basapur (2), Majjigudda and Bhadrapur as has been already discussed, the low resistivity layers of 7, 3.3 and 1.4 Ohm-m, infer the occurrence of brackish water zones. But the practical experience shows that the fresh water was struck in the bore wells without encountering brackish water zone at any level. At the same time, in case of Padesur and Navalgund, though the interpreted layer parameters clearly give the of the occurrence of brackish water, in the 2nd layer whose resistivity indication is between 1 to 2 Ohm-m. However, these layers are followed by a layer of 25 Ohm-m resistivity which is approximately 2 meters thick (Padesur), indicating fresh water zones. The drilling results conclude that these two are brackish water borewells. Hence due to above mentioned complication and drawbacks the inferences are not always certain and the recommendations are at risk. Resistivity interpretation can hence become more dependable solution for delineating the fresh and brackish water aquifers if it can be supported by some easy to observe and resolve type of technique. Such a technique and solution can be visualised in the qualitative behaviour of the apparent resistivity data in the field while carring out resistivity investigations.

Qualitative field data analysis

An easily comparable, convincing and interesting contrast of the apparent resistivity values for soundings are observed for fresh water and brackish water borewells. The apparent resistivity data of the soundings carried out at the fresh water borewells and brackish water borewells are given in Table-4 along with the interpreted layer parameters. The qualitative data for the inter electrode spacing (a) between 15 and 100 m for all the soundings analysed is in Table 4. The apparent resistivity values for the sounding at Basapur (1), Majjigudda and Bhadrapur, range from 20 to 50 Ohm-m, 16-45 Ohm-m, 25-98 Ohm-m and 15-40 Ohm-m respectively as shown (Table 2). For the brackish water sites the apparent resistivity values for Padesur and Navalgund, range from 3 to 11 Ohm-m and 4 to 15 Ohm-m, respectively. The averages range of apparent resistivity values for fresh water borewells is 15 to 90 Ohm-m and that of brackish water borewells is 3 to 15 Ohm-m in the study area. From these ranges, it is interesting to observe that the lowest values apparent resistivity of the brackish water range is 3

Ohm-m and the fresh water aquifer is 15 Ohm-m and the highest values of apparent resistivity range for brackish water aquifer is 15 Ohm-m and for fresh water region is 90 Ohm-m. It is very clear that they are conspicuous and almost unique, as they are not overlapping, as far as the study area is concerned. These ranges do not mix with each other and hence become independent signatures of the subsurface resolution of groundwater quality. The range of apparent resistivity 3 to 15 Ohm-m for brackish water zones is considerably low when compared the range of 15 to 90 Ohm-m for fresh water zones. Hence it facilates an easy and unbaised resolution of the geophysical investigation into two contrasting groups as

(a) The sites which record relatively very low apparent resistivity values and yield brackish ground water as in the case of Navalgund and Padesur.

(b) The sites which record relatively very high apparent resistivity values when compared with apparent resistivity values of (a), and yield fresh ground water as in the case of Basapur (1) and (2), Majjiguda and Bhadrapur.

Discussion

The field apparent resistivity values invariably stick to their behaviours by reflecting distinguishable and almost independent non overlapping apparent resistivity domains for fresh water and brackish water sites. The brackish water sites record relatively low apparent resistivity values for the electrode separation 30-100 m and on the contrary the fresh water sites record relatively

higher apparent resistivity values compares to the lower values of brackish water for the same electrode separation. The technique is totally cost free and simple. It is a biproduct of the routine resistivity surveys which are the basic requirements for delineating of fresh and brackish water zones. The interpretation of the qualitative field data, can be made in the field itself to have a confidant first hand information.

In the hybrid technique, first the field resistivity data at known brackish water sites and fresh water sites in the study area should be critically analysed and the apparent resistivity domains for the brackish and the fresh aquifers must be deduced. This can be taken as a criteria to resolve fresh water sites and brackish water sites. The inferences of the quantitative interpretation should be compared with the inferences of qualitative field data technique and the ambiguities, risks and uncertainties should be tackled. When the quantitative and qualitative inferences agree with each other, the solution becomes crystal clear with double confidence. On the other hand if their is a disagreement between the qualitative and quantitative results due to the difficult to resolve the deceptive layer parameters by virtue of their overlapping resistivity ranges, the qualitative results which are more invariable, compared to the interpretated layer parameters in their resolution can be taken into consideration, in order to deal the risky situation of recommendations. Such a comparision facilates in arriving at an unbaised interpretation of the subsurface situation which will be more confident and dependable.

Table –1: Qualitative and quantitative resistivity data of the sounding curves discussed in the paper

Name of the Village	Interpretated layer parameters of the soundings		Range of apparent resistivity	Corresponding inter-electrode spacing, a	Quality of groundwater
	(Ohm mtrs)	(m)	(Ohm mtrs)	(m)	
Basapur(1)** (W)	6.39 22.48 154	1.18 23 …	20-50	a = 12-60	Fresh and potable
Basapur(2)** (W)	70 7	0.9 13.5	16-45	a = 16-45	-- do –

Name of the Village	Interpreted layer parameters (Ohm mtrs)	(m)	Range of apparent resistivity (Ohm mtrs)	Corresponding inter-electrode spacing (m)	Quality of groundwater
	High	--			
Majjigudda** (W)	4 1.0 High	1.66 1.35 --	25-98	a = 20-100	-- do --
Bhadrapur** (W)	20 3.3 4.84 High	0.49 14.3 --	15-40	a = 35-80	-- do --
Padesur◆ (S)	3.9 1.78 1.79 99830	1.07 16.48 --	3-11	a = 20-80	Brackish
Navalgund◆ (W)	14.13 2.12 12.67 High	0.23 11.98 3.98 --	4-15	a = 20-60	-- do --
Saidapur* (W)	6.71 4.68 44.51 --	1.97 4.09 4.96	25-63	a = 3-100	The site yields 1500 GPH of fresh water
Shelwadi* (W)	3.14 4.55 High	2.38 12.83 --	13-45	a = 30-100	The site yields 250 GPH of fresh water
Kumargoppa* (S)	17.83 3.7 40.59 −	1.72 1.25 63.65	20-55	a = 30-100	The site yields 2000 GPH of fresh water
Sottakanal* (W)	6.01 10.01 14.62 23.28 50.00 332.82	7.28 3.69	21-49	a = 30-100	The site yields brackish water

----

** Existing fresh water borewell   *Site recommended by using Hybrid Technique
◆ Existing brackish water borewell   W: Wenner Config   S: Schlumberger Config.

Table-2: Interpretated results of soundings near fresh water borewells

Name of the Village	Interpreted layer parameters of the sounding (Wenner Configuration) (Ohm mtrs)	(m)	Range of apparent resistivity (Ohm mtrs)	Corresponding inter-electrode spacing (m)	Quality of groundwater
Basapur(1)	6.39 22.48 --	1.18 23	20-50	a = 12-60	Fresh and potable
Basapur(2)	70 7	0.9 13.5	16-45	a = 16-45	-- do --

	High	--			
Majjigudda	4	2.3	25-90	a = 20-100	-- do --
	1.6				
	High	--			
Bhadrapur	20	0.49	15-40	a = 35-80	-- do --
	0.35				
	4.84	14.3			
	High	--			

Table–3: Interpretated results of soundings at brackish water borewels

Name of the Village	Interpretated layer parameters of the sounding (Wenner Configuration) (Ohm mtrs)	(m)	Range of apparent resistivity (Ohm mtrs)	Corresponding inter-electrode spacing (m)	Quality of ground-water
Padesur	3.9	1.07	3-11	a = 20-80	Brackish
	1.78	16.48			
	24.44	1.79			
	99830	--			
Navalgund	14.13	0.23	4-15	a = 20-60	-- do --
	2.12	11.98			
	12.68	3.98			
	8.0	--			

Table-4: Apparent resistivity data and interpretated results of fresh and brackish water borewells

Fresh water borewells					Bbrackish water borewells				
Location	Interelectrode Separation(a) (Wenner Conf)	Range of apparent resistivity	Interpretated results Ohm-m	m	Location	Interelectrode separation(a) (Wenner Conf)	Range of apparent resistivity	interpretated results Ohm-m	m
Basapur(1)	a= 12-60	20-50	6.39	1.18	Padesur	a= 20-80	3-11	3.9	1.07
			22.48	23.6				1.78	16.48
			154.80	--				24.43	1.79
								High	--
Majjigudda	a= 20-100	25-98	4.45	1.66	Navalgund	a= 20-60	4-15	14.13	0.23
			1.1	1.35				2.12	11.98
			High	--				12.67	3.98
								High	--
Bhadrapur	a= 35-80	15-40	20.0	0.49					
			3.33	0.35					
			4.84	14.31					
			High	--					

194

Table-5: Details of reccommendation and drilling results (basedon hybrid technique)

Name of the Village	Interpretated layer parameters of the sounding (Wenner Configuration) (Ohm mtrs)	(m)	Range of apparent resistivity (Ohm mtrs)	Corresponding inter-electrode spacing (m)	Quality of groundwater
Saidapur*	6.71	1.97	25-63	a = 3-100	The site yields
	4.68	4.09			1500 GPH of
	44.51	4.96			fresh water
	137.24	--			
Shelwadi*	3.14	2.38	13-45	a = 30-100	The site yields
	4.55	12.83			250 GPH of
	High	--			fresh water
Kumargoppa*	17.83	1.72	20-55	a = 30-100	The site yields
	3.7	1.25			2000 GPH of
	40.59	63.65			fresh water
	1801	–			
Sottakanal*	6.01	7.28	21-49	a = 30-100	The site yields
	10.01	3.69			brackish water
	162.93	14.62			
	25.97	23.28			
	87.03	50.00			

Based on the above said criteria four sites were recommended for drilling of borewells in the study area. The sites are Saidapur, Shelwadi, Kumargoppa and Sottakanal. The interpreted layer parameters and the range of apparent resistivity values for a range of electrode seperation, recommended depth of drilling. The results are given in table-5. The sounding curves are shown in figures 7 and 8.

The study at Saidapur favours the occurrence of fresh water beneath. The layer parameters and the qualitative interpretation agree well. The apparent resistivity values for a = 30-100 m are in the range of 25-63 Ohm-m. At the same time the III layer occurs at a depth of 6 m and is 4 m thick. Its resistivity is 44 Ohm-m, followed by a harder formation. The above results agree well and hence the recommendation was approved. The site yields 1500 GPH of fresh water. In case of Shelwadi and Kumargoppa the results are difficult to resolve. The field data analysis signals the occurrence of fresh water. On the other hand the low resistitity layers of 4 Ohm-m at Shelwadi is 12 m thick and interpretes the brackish water zone. At Kumargoppa another layer whose resistivity is 40 Ohm-m infers the presence of fresh water. On the whole it becomes difficult to take a confidant decision. The sites where recommend for drilling based on the quality of field data. Both the sites yield fresh water. However, the site at Sottakanal also posses problems of the type explained above and yield brackish water. Please see table-4 for clarification. The results indicate fairly successfull performance of the field data analysis in fruitfully supporting the resistivity data interpretation. Three out of the four cases yield fresh water. The nature and behaviour of the apparent resistivity values seems to be more invariable compared to the interpreted resisitivity data. Hence a combination of both the analysis proves to be a more realistic and believable resolution for recommending sites for exploitation of fresh ground water acquifers in brackish terrains.

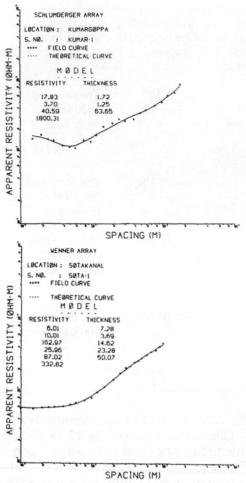

Figure 8: Resistivity sounding interpretation at Kumargoppa and respectively

Conclusion

Resistivity data interpretation sometimes posses the problems of uncertainty, while resolving the fresh and brackish water aquifers. The analysis of the field resistivity data provides a confidant and realistic support in the resolution and facilates considerably in avoiding the risks and uncertainties while recommending sites for exploitation of groundwater in quality affected regions.

Referances

Hodlur, G.K., Deshmuck , S.D., amachandr,Y.M., Rao ,.G. and Gupta, C.P.. Geophysical investigations for exploration of fresh water pockets in regionally brackish aquifers of Navalgund Taluk, Dharwar dist., Karnataka. Technical Report No. 111 NGRI-ENVIRON.

Lazreg H., 1972, Master curves for Wenner arrays. Inland water Director Water Resources branch; Ottava, Canada, pp 109.

Orllana E, Mooney H.M.(1966) Master tables and curves for vertical scounding over layered structure interciencia.

Ozoff, K. and Jupp, D.L.B., 1975 Joint inversion of geophysical data Geophys. J. R. Astr. Soc, Pp977-991.

Verma, S.K. and Panthulu, K.P., 1990, Software for the interpretation of resistivity sounding data , NGRI, monograh,pp.120

Intl. Conf. on Sustainable Development and Management of Groundwater Resources
in Semi-Arid Region with Special Reference to Hard Rock, (IGC 2002),
M. Thangarajan, S.N. Rai & V.S. Singh (Eds.)

Hydrogeochemistry and quality assessment of groundwater in and around Salem magnesite mine area, Salem District, Tamil Nadu

M.Sathyanarayanan and P.Periakali

Department of Applied Geology, University of Madras, Guindy Campus, Chennai-600025.

Abstract

The chemical characteristics of groundwater in and around Salem magnesite mine area, Salem district, Tamil Nadu, have been studied to evaluate the suitability of water for irrigation and domestic uses. Fifty water samples representing the deep and shallow groundwater of the area were collected and analysed for pH, Ec, TDS, Ca, Mg, Na, K, HCO3, CO3, SO4, Cl and minor/trace elements Fe, Mn, Ni, Cr and Co. It was observed that the quality of groundwater is not suitable for domestic use with some exception. The observed high quantity of trace elements at some sites requires immediate attention and detailed study. Lateritic weathering in the upper weathered and decomposed/disintegrated rock appears to have mobilised and depleted Ni, Cr and Co from the mine overburden and ultrabasic rocks enriching the same in groundwater. The calculated value of SAR, RSC and %Na indicate the good to permissible quality of water during both summer and winter season.

Keywords: Salem, hydrogeochemistry, groundwater pollution, magnestie mine

Introduction

Groundwater is the only replenishable natural resource available to man. While availability of water is one side of the problem, the deteriorating quality of existing water resources is the other side causing great concern. All groundwater contains minerals carried in solution, the type and concentration of which depends upon the surface and subsurface environment, rate of groundwater movement and source of groundwater (Walton, 1970; Prasad, 1984). Man can adversely alter the chemical quality of groundwater by permitting highly mineralised water to enter into fresh water through mining activity. While pollution of groundwater due to external contaminants such as industrial, urban and agricultural activities is quite well documented, one of the aspect on which enough attention has not been focussed is the degradation of groundwater quality caused by opencast mining activity. An attempt has been made therefore to evaluate the quality of groundwater by collecting fifty water samples from tube wells and dug wells in and around Salem magnesite mine area, Salem district and to assess the suitability and causes for deterioration of groundwater quality in this region.

Study area

The study area is confined between latitudes 11°37'20"N and 11°46'N and longitudes 78°05'E to 78°11'20"E in parts of the Survey of India toposheet nos. 58 I/1 and 58 I/2, covering an area of about 193 sq.km. It is surrounded by an amphitheater of high releif area in the northern part while in the southern part with residual and denudational hills and undulating terrain. The general elevation ranges between 250 and 320 meters, and the hill ranges attain a height of about 1,200 to 1,500 meters.

The study area has a mining history known for chromite and magnesite associated with mining related environmental hazard. Records and reports indicate about ancient chromite mining activity in the area and at present only imprints and scars in the form of wet weather rills are seen. With a view to understand the impact of active and abandoned mine lands due to magnesite mining and the subsequent environmental consideration, this classic area has been chosen for this research study.

Geologically, the terrain is comprised of the peninsular gneiss, charnockite, basic granulites, ultramafic complex and potassic members, and bands of magnetite quartzite (Fig.1). Magnesite of Salem (Chalk hills) area is closely associated with the chromiferrous ultrabasic series occurring as intrusive masses amidst highly metamorphosed rocks that have been subjected to repeated period of deformation.

Geomorphologically, the terrain is developed out of pedimentation and peneplanation, processes characteristic of sub-tropical regions. The only prominent structural dome (Nagaramalai) is seen in the form of a tabloid plug like ultramafic intrusive body. The river Thirumanimuttar bisects the terrain with its minor tributaries flowing in the NE-SW direction.

Structurally, the charnockite series and peninsular gneiss have been subjected to repeated deformation and the last deformation related to pink granite emplacement, has completely obliterated the structural patterns. Owing to the emplacement of pink granite, the earlier rocks have been folded in the form of overturned antiform with a NE-SW trend and northeastern plunge. Dunite is characterised by joints in random direction. The N-S shear zone, NE-SW and NS-EW trending lineaments in the southeastern part of the study area have been inferred and identified from imagery and groundtruth. Rainfall is seasonal in the area contributed by south-west and north-east monsoon between June and December. The total rainfall ranges between 800-1300mm. Evapotranspiration is greater than precipitation for a minimum of five months and the main river and its minor tributaries are non-perennial.

The normal mean annual temperature of the region varies between 20 and 35°C. The temperature increases up to 38°C and, with the onset of monsoon in June, there is considerable drop in temperature. April and may are the hottest months with a mean daily maximum temperature of about 38°C and January is the coldest month with a mean minimum temperature of 18°C. The estimated mean temperature of the soil from the surface to a depth of 110 cm exceeds that of the atmosphere by about 3.4°C.

Methodology

A total 50 representative water samples (41 tube wells + 9 dug wells) were collected from various locations in and around Salem magnesite mine area (Fig.2) during summer and winter of 1999. For sampling, wells which are in constant use and approachable were selected. Pre-washed one litre polyethylene bottles were used for sample collection and preservation. pH and electrical conductivity (EC) were measured at the sampling sites. Conductivity and pH were measured using portable Elico EC and pH meter. In the laboratory the water samples were analysed for the cations (calcium, magnesium, sodium and potassium) and anions (carbonate, bicarbonate, chloride and sulphate). Analysis of Ca, Mg, Na and K were carried out by titrimetric and flame photometric method. Replicates were run for each sample for cation analysis and the blanks were checked after every 15 samples. Concentration of Cl was measured by argentometric titration. The amount of carbonate and bicarbonate was determined by Nephelometer (APHA, 1995). Overall reproducibility for major ions was within +/- 5%. Cationic and anionic charge balance (<10%) is an added proof of the precision of the data. For determining the trace metal content 250 ml water samples were collected seperately from all fifty locations. These samples were acidified with nitric acid immediately. The trace metals (Fe, Mn, Ni, Cr, Co) in these samples were analysed by using AAS, following analytical methods described by Brown et al. (1974). The optimum working range of various trace metals were 1-100 µg/l (Fe), 0.5-60 µg/l (Mn), 0.1-20 µg/l (Ni), 0.06-15 µg/l (Cr) and 0.05-15 µg/l (Co).

Results and Discussion

Hydrochemistry

The analytical results of the physico-chemical analysis of dug well and tube

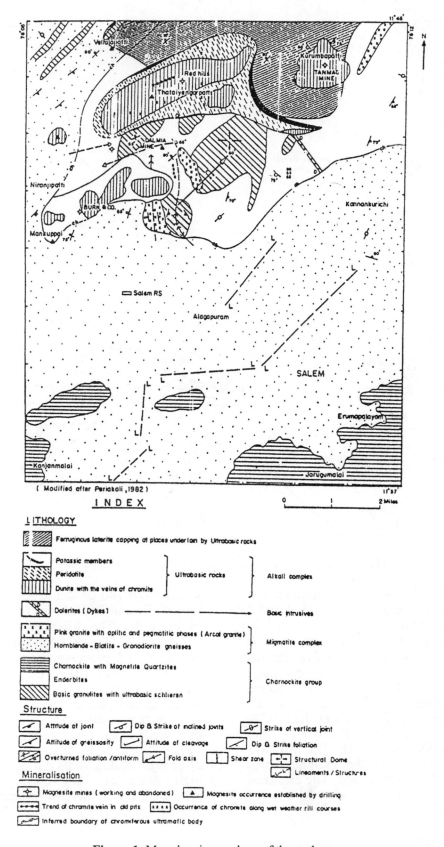

(Modified after Periakali ,1982)

INDEX

LITHOLOGY

	Ferruginous laterite capping at places underlain by Ultrabasic rocks

	Potassic members		
	Peridotite	Ultrabasic rocks	Alkali complex
	Dunite with the veins of chromite		

	Dolerites (Dykes)	Basic intrusives

	Pink granite with aplitic and pegmatitic phases (Arcot granite)	Migmatite complex
	Hornblende – Biotite – Granodiorite gneisses	

	Charnockite with Magnetite Quartzites	
	Enderbites	Charnockite group
	Basic granulites with ultrabasic schlieren	

Structure

	Attitude of joint		Dip & Strike of inclined joints		Strike of vertical joint
	Attitude of greissosity		Attitude of cleavage		Dip & Strike foliation
	Overturned foliation /antiform		Fold axis		Shear zone		Structural Dome
					Lineaments / Structures

Mineralisation

	Magnesite mines (working and abandoned)	Magnesite occurrence established by drilling
	Trend of chromite vein in old pits	Occurrence of chromite along wet weather rill courses
	Inferred boundary of chromiferous ultramafic body	

Figure 1: Map showing geology of the study area

199

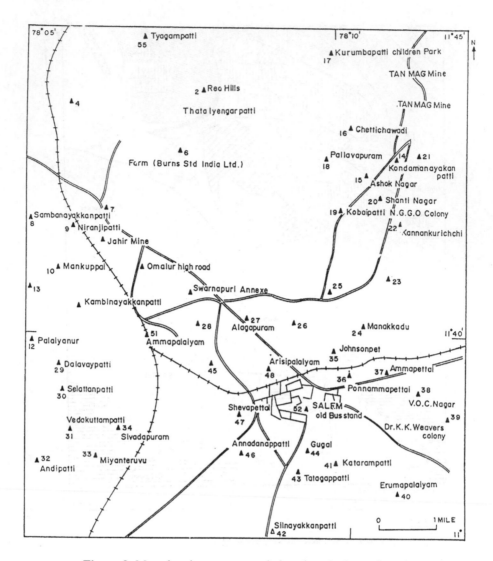

Figure 2: Map showing water sample locations in the study area

well waters collected during summer and winter seasons are given in tables 1 and 2 respectively. During summer, the pH of the analysed samples varies from a low of 5.7 for the sample obtained at Tyagampatti (55) to a higher value of 9.6 at Niranjipatti (9), indicating slightly acidic to alkaline nature of water samples. During winter, the pH varies from 7.0 at V.O.C nagar (38) to 8.5 in between Palaiyanur and Kambinayakkampatti (13). The pH was generally high in the mining area throughout the study period. The electrical conductivity (EC) values varied from 422 μS/cm to 3154 μS/cm for tube wells and 1079 μS/cm to 1980 μS/cm in dug wells during summer. In winter, the EC values

varied from 740 μS/cm to 4261 μS/cm for tube wells and 813 μS/cm to 3143 μS/cm in dug wells. The differences between the value may reflect the wide variation in the activities and processes prevailing in the region. The total dissolved solids (TDS), which is the sum of the dissolved ionic concentration varied between 350-9500 mg/l in tube wells and 550-3000 mg/l in dug wells during summer, and between 1092-3900 mg/l in tube wells and 1170-2652 mg/l in dug wells during winter. The high value of TDS during winter is attributed to the intense weathering and subsequent leaching of minerals from the exposed surrounding rocks.

The anion chemistry shows that bicarbonate is the dominant ion both in deep and surface water aquifers. In the water of deep water (tube wells) the concentration of bicarbonate varies from 186 mg/l near Dalmia magnesite factory (6) to 995 mg/l at Miyanteruvu (33) during summer and from 109 mg/l at Silnayakanpatti (42) to 700 mg/l at Niranjipatti (9) during winter. In waters of shallow aquifer, the concentration ranges between 334 to 873 mg/l during summer and 174 to 622 mg/l during winter. The $CO_3 + HCO_3$ togeather contribute on an average 44% and 56% of the total anions both in tube wells and dug wells respectively during summer, 36% and 46% during winter in equivalent units. Bicarbonates derived mainly from the soil zone CO_2 and at the time of weathering of parent minerals. The soil zone in the subsurface environment contains elevated CO_2 pressure (produced as a result of decay of organic matter and root respiration) which in turn combines with rain water to form bicarbonate:

$$CO_2 + H_2O == H_2CO_3$$
$$H_2CO_3 == H^+ + HCO_3$$

Bicarbonate may also be derived from the dissolution of carbonates and/or silicate minerals by the carbonic acid from the reactions:

$CaCO_3 + H_2CO_3 == Ca^{2+} + 2HCO_3^-$
(Limestone)
$CaMg(CO_3)_2 + 2H_2CO_3 == Ca^{2+} + Mg^{2+} + 4 HCO_3^-$
(Dolomite)
$2NaAlSi_3O_8 + 2H_2CO_3 + 9H_2O == Al_2S_2O_5(OH)_4 +$
(Albite) $2Na^+ + 4H_4SiO_4 + 2HCO_3$
(Kaolinite)

The high concentration of bicarbonate indicates, intense chemical weathering in the area. The bicarbonate in groundwater is mainly due to dissolution of magnesite minerals by rain and surface waters. The chloride concentration during summer varies between 10-435 mg/l in tube wells and 42-682 mg/l in dug wells. Abnormal concentration of chloride may be due to pollution by sewage wastes and leaching of saline residues in the soil. The concentration of sulphate varies from 9-758 mg/l in tube wells and 7-109 mg/l in dug wells during summer and from 7-604 mg/l in tube wells and 20-119 mg/l in dug wells during winter. The sulphate is usually derived from the oxidative weathering of sulphide bearing minerals present in the area.

The major cations include Ca, Mg, Na and K. The cationic chemistry is dominated by sodium and magnesium ions during summer and, magnesium and calcium during winter. The concentration of sodium in the tube well sample was in the range of minimum of 22 mg/l near Dalmia magnesite factory (6) and maximum of 467 mg/l at V.O.C nagar (38) during summer. The concentration of the same ion in dug wells varies from a minimum value of 83 mg/l to a maximum of 284 mg/l. On an average sodium and potassium account for 49% and 51% of the total cations (TZ^+) in tube wells and dug wells respectively during summer. The order of cationic abundance is given as sodium followed by magnesium, calcium and potassium (Na>Mg>Ca>K) in tube well and dug well during summer. The predominance of (Na+K) over magnesium and calcium is attributed to silicate weathering particularly due to the presence of orthoclase and biotite of peninsular gneiss in the study area. The low value of potassium than sodium is due to greater resistance to weathering of the potassium and its fixation in the formation of clay minerals.

During winter the concentration of magnesium in tube well samples was minimum of 44 mg/l at Silnayakanpatti (42) and maximum of 367 mg/l at Selatampatti (30). In dug wells it varies from a minimum value of 48 mg/l to a maximum of 228 mg/l. On an average, magnesium and calcium account for 84% and 81% of the total cations (TZ^+) in tube wells and dug wells respectively. The order of cationic abundance is given as magnesium followed by calcium, sodium and potassium (Mg>Ca>Na>K) in tube wells and dug wells during winter. The predominance of magnesium is attributed to the influence of magnesite mining activity and subsequent leaching due to percolation of rain water through the soil and weathered products.

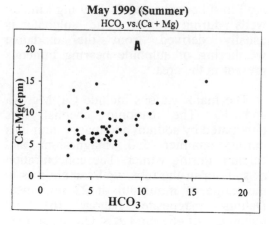

May 1999 (Summer)
HCO₃ vs.(Ca + Mg)

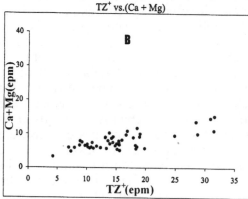

TZ⁺ vs.(Ca + Mg)

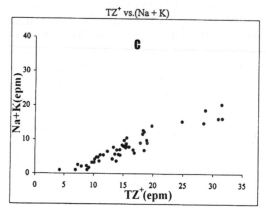

TZ⁺ vs.(Na + K)

Figure 3: Scatter diagram between (A) (Ca+Mg) vs. HCO₃, (B) (Ca+Mg) vs. Total cations (TZ⁻¹) and (C) (Na+K) vs. Total cations (TZ⁻¹) during summer 1999.

The abundance of various ions can be modelled in terms of weathering of various rock forming minerals (Singh and Hasnain, 1998, 1999). However, the solution products of silicate weathering are difficult to quantify because of the degradation of silicates incongruently generates a variety of solid phases (mostly clays) along with dissolved species. The plot of (Ca+Mg) vs.HCO3 shows that the data points in both tube wells and dug wells during summer and winter fall above the equiline towards (Ca+Mg) side (Fig.3A & 4A). This situation requires neutralisation of alkalis by carbonate alkalinity. The (Ca+Mg) vs.TZ⁺ plot lies below the equiline for dug well as well as bore well during summer and winter (Fig.3B & 4B). The relative high concentration of (Na+K) to the total cations indicates that silicate weathering and/or contributions from alkaline soils act as the main sources of major ions to these waters (Fig.3C & 4C) and may be due to weathering of plagioclase feldspar in peninsular gneiss.

Trace metals

Lateritic weathering which covers the ultramafic rock complexes of the area are subdivided into autochthonous and semiautochthonous weathering horizons. The latter are constituted by a mixture of a) autochthonous Fe-Cr rich limonitic material derived from ultramafic rocks and b) Al-Si-Ti-Zr rich weathering material, derived from spatially associated sialic rocks (Marker, 1988). In the upper weathered and decomposed/disintegrated rock, Cr, Mn, Co and Ni are subjected to mobilisation and depletion (Marker et al, 1989), which is of considerable importance in the enrichment of these trace elements in soil and water. To understand the interaction of ultramafic rocks with water, trace elements Ni, Cr and Co are analysed along with Fe and Mn.

The concentration of iron in groundwater exceeded the permissible limit (0.3 mg/l) in about 67% of the samples in summer and 59% of the samples in winter. Higher concentration of iron was observed towards the southern part of the area, particularly at Silnayakanpatti, Vedakuttampatti, Shevapettai, Kalarampatti and Ammapettai. The western and northern parts near Mankuppai and Tyagampatti also indicate higher values. The presence of dissolved iron appears to be influenced by the subsurface lithology and close as proximity to source of contamination.

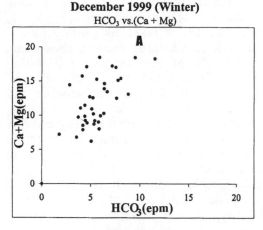

December 1999 (Winter)

HCO₃ vs.(Ca + Mg)

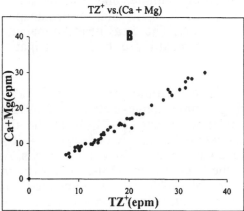

TZ⁺ vs.(Ca + Mg)

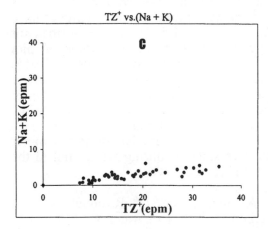

TZ⁺ vs.(Na + K)

Figure 4: Scatter diagram between (A) (Ca+Mg) vs. HCO₃, (B) (Ca+Mg) vs. Total cations (TZ^{-1}) and (C) (Na+K) vs. Total cations (TZ^{-1}) during winter 1999.

Manganese concentration exceeded the permissible limit (0.1 mg/l) in about 24% of samples in summer and 11% of samples in winter. Agricultural practices (fertilizer use, sewage and animal waste disposal), and atmospheric deposition from fossil fuel combustion and municipal incinerators contribute significant amount of manganese. The distribution of manganese is similar to iron in groundwater of the study area. Higher concentration of manganese was recorded in southern part of the study area, particularly in Dalavaypatti, Vedakuttampatti, Shevapettai, Ammapettai, Silnayakanpatti, Gugai and near Kannankurichi.

Nickel exceeded the permissible limit (0.1 mg/l) in about 63% of the samples in summer and 7% of samples in winter. Higher concentration was observed in some pockets and particularly in areas around Selatanpatti, Miyanteruvu, Sivadapuram, Alagapuram, Annadanapatti, Ammapettai, Katarampatti, Karuppur (West) and near Palykaradu.

Chromium in groundwater exceeded the permissible limit (0.1 mg/l) in about 63% of the samples in summer and 62% in winter. It was observed that most of the northern and northeastern parts are contaminated by chromium. Thathiangarpatti, Red Hills, Karuppur (west), Jagir mines, Manakkadu and Kannankurichi show higher concentration of chromium in groundwater. The source of chromium is from weathering and subsequent leaching of ultrabasic rocks.

Cobalt exceeded the permissible limit (0.1 mg/l) in about 14% of samples in summer and was below permissible limit in winter. The southern part of the area shows high cobalt concentration in groundwater, when compared to northern and central portions particularly in Salem (proper), Silnayakanpatti, Miyanteruvu, Annadanapatti, Katarampatti, Tatagapatti, Shevapettai, Arisipalayam, Ammapalayam, Tyagampatti and TANMAG mines.

Quality Assessment

The data obtained by chemical analyses were evaluated in terms of its suitability for drinking and irrigation purpose. Table 3 shows the range of ionic concentrations in the tube well and dug well water samples during summer and winter, and the maximum permissible limit prescribed by WHO (1996) and ISI (1995). Analytical

203

data shows that the groundwater of the area are not suitable for drinking and domestic uses with few exceptions. The values of TDS and EC exceeded the permissible limit at some sites indication the higher concentrations. The total hardness (TH) varies between 46-221 mg/l and 85-160 mg/l during summer indicating moderate to hard type of water. During winter the TH value varies between 104-495 mg/l in tube well and 93-404 mg/l in dug well indicating moderately hard to very hard type of water. It was observed that a major portion of the area is characterised by waters of carbonate hardness in both seasons. Temporary hardness is due to the interaction of water (meteoric and groundwater) with magnesite and magnesite bearing rocks exposed due to mining.

The parameters such as sodium adsorption ratio (SAR), percent sodium (%Na) and residual sodium carbonate (RSC) were computed to assess the suitability of water for irrigation purposes. The total concentration of soluble salts in irrigation water can be expressed for the purpose of classification of irrigation water as low (EC =<250 μS/cm), medium (250-750 μS/cm), high (750-2250 μS/cm) and very high (2250-5000 μS/cm) salinity zone (Richards, 1954). While a high salt concentration in water leads to formation of saline soil, a high sodium concentration leads to development of an alkaline soil. The sodium or alkali hazard in the use of water for irrigation is determined by the absolute and relative concentration of cations and is expressed in terms of sodium adsorption ratio (SAR).

$$SAR = [Na/(Ca+Mg)/2]^{0.5}$$

There is a significant relationship between values of irrigation water and the extent to which sodium is adsorbed by the soils. If water used for irrigation is rich in sodium and low calcium, the cation-exchange complex may become saturated with sodium. This can destroy soil structure owing to dispersion of the clay particles. The calculated values of SAR in the area ranges from 0.53 to 8.66 in the tube wells and 2.74 to 7.30 in the dug wells during summer and 0.15 to 1.44 in the tube wells and 0.58 to 1.74 in the dug

wells during winter. The plot of the data on the US salinity diagram (Fig.5), in which the EC is taken as salinity hazard and SAR as alkalinity hazard, shows that nearly 90% of the samples fall within C3S1, C3S2, C2S1 and C4S2 classes. About 53% to 73% of the samples fall within C3S1 class in both seasons, indicating medium to high salinity and low alkalinity, which can be used for irrigation on almost all soils, with little danger of the development of harmful level of exchangeable sodium. It is moderate to unsuitable, however, at Chettichavadi, Dalavaypatti, Miyanteruvu, Ponnammapettai during summer and in Selatanpatti, Tatagapatti, Gugai and Annadanapatti during winter. The areas near Ammapettai and Katarampatti are bad for irrigation during both seasons.

The sodium percentage (%Na) during summer ranges between 12-69% in the tube wells and 26-71 % in the dug wells, while during winter between 4-26% in tube wells and 12-30% in dug wells. As per the Indian standards, a maximum sodium of 60% is recommended for irrigation water. In the study area, eight out of forty one tube well water samples and one out of nine dug well samples have higher values of %Na than prescribed for irrigation purposes during summer. A plot of analytical data on a Wilcox (1955) diagram relating electrical conductivity and sodium percent shows that water of the study area is good to permissible quality for irrigation purpose (Fig.6). However, three samples during summer and five samples during winter fall in the unsuitable category.

A high value of Residual Sodium Carbonate (RSC) in water value leads to an increase in the adsorption of sodium on soil (Eaton, 1950). Irrigation waters having RSC values greater than 5 meq/l have been considered harmful to the growth of plants, while waters with RSC values above 2.5 meq/l are generally considered unsuitable for irrigation purpose. The RSC values between 1.25-2.5 meq/l is of marginal quality, while value less than 1.25 indicates that water is probably safe. Based on the values, 80% of the groundwater samples have values less than 1.25 and are safe for irrigation during both the seasons. Only 7 % of the samples exceed the RSC value of 2.5 in areas around Chettichavadi and Niranjipatti rendering the groundwater unsuitable for irrigation during summer and winter.

predominant type of hydrochemical facies among the cationic phase but the dominance of bicarbonate type of waters was observed among anions. However, during winter Mg-Ca-HCO3 was observed to be the dominant hydrochemical facies.

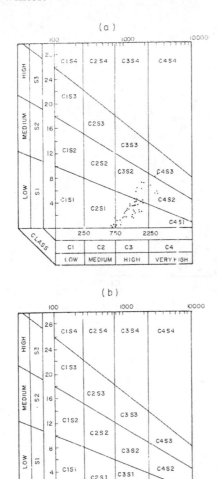

Figure 5: USSL classification of groundwater during (a) summer and (b) winter, 1999

The evolution of water and relationship between rock types and water composition can be deciphered from the trilinear Piper (1944) diagram (Fig.7). The plot of chemical data on trilinear diagram reveals that the water samples fall in the areas of 1,2,3,4,5,7 and 9 during summer and in the areas of 1,3,5 and 6 during winter, indicating chemical characters of the water as given in the table 4. During summer water samples exibit no

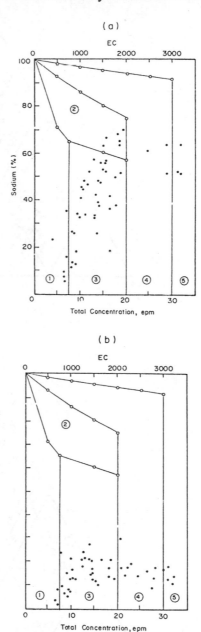

Legend

1. Excellent to good 2. Permissible to doubtful 3. Good to permissible 4. Doubtful to unsuitable 5. Unsuitable

Figure 6: Wilcox's diagram for classifi- cation of irrigation waters during (a) summer and (b) winter, 1999 (after Wilcox, 1955)

205

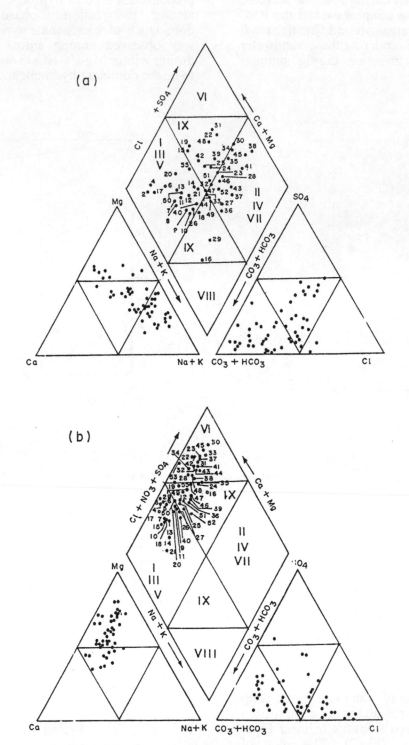

Figure 7: Trilinear diagram for ground-water during (a) summer and (b) winter, 1999

Concluding Remarks

Results of the entire study has enabled to identify the separation of natural and anthropogenic sources and to determine the feasibility of various mitigation strategies posed due to the magnesite mining activity. As there is no major magnesite occurrence in the southern part of the study area, all relevant parameters obtained forms the "baseline data" to compare with that of the northern part of the area where intense magnesite mining activity and man induced environmental hazards are taking place. The area between 11°37' - 11°41'N can be taken for reference as baseline data and the area between 11°41'-11°46'N as problem oriented due to man induced mining hazard and environment.

The results also underscore the crucial need for sound scientific information in predicting, assessing and remediating the enviromental effects of magnesite mining. The effect of chromium has become a long term environmental concern for the habitants dwelling south of the mining activity area and in the urbanised Salem town. Extensive remedial effects are required to isolate toxic trace elements (Ni, Cr and Co) from the spoils/dumps and other soluble metals.

The main problem faced in the mining hazard environment is the need for safe dumping sites/land for spoils, secure landfills to prevent water percolating into the already chromite bearing lithological entity/domain.

A preliminary regional magnesite mining hazard study programme can be formulated on existing geochemical data information about mineral deposits, geology, ecoregion zones, hydrology and land use. Generalization of contamination for degradation process can be developed based on the new information gained which will enable people involved in regional, national and international agencies in the preparation of environmental impact assessment map. It is necessary to create public awareness regarding the deterioration of groundwater quality and its long term effect on health of human beings, birds and animals.

References

APHA (American public Health Association) (1996) Standard methods for the Examination of warter and wastewater, 19th eds. Public Health Association,Washington,DC.

Brown, E., Skougstad, M.W. and Fishman, M.J. (1974). Methods for collection and analysis of water samples for dissolved minerals and gases. U.S.Dept. of Interior, Book-5, 160p.

Eaton, E.M. (1950) Significance of carbonate in Irrigation Water. Soil Science, v.69, pp.123-133.

ISI (Indian Standards Institution) (1995) IS:10500-1995. Specification for drinking water. Indian Standards Institution, New Delhi.

Marker, A. (1988) Lteritische Vermitterungsdecken uber ultramafischen Gesteinskomplexen in Brasilien und den Philippinen. Unpublished Ph.D. thesis, Institute of Mineralogy, RWTH Aachen. 319pp.

Marker, A., Friedrich, G. and Carvalho, A. and Meifi, A. (1989) Control of the distribution of Mn, Co, Zn, Cr, Ti and REE during the evolution of lateritic weathering covers above ultramafic rock complexes. Abstracts of XIII International Geochemical Exploration Symposium & II Brazilian Geochemical congress, Rio de Janeiro, Brazil, Oct.1989. pp.30-31.

Piper A.M (1944) A graphic procedure in the geochemical interpretation of water analysis. Trans Amer Geophy Union, v25. pp 914-928.

Prasad N.B.N (1984). Hydrogelogical studies in the Bhadra River Basin, Karnataka, Unpublished Ph.D., thesis (University of Mysore), 323 p.

Richards, L.A. (1954). Diagonosis and improvement of saline and alkali soils. U.S. Dept. Agri. Hand Book, No.60, 160p.

Singh, A.K. and Hasnain, S.I. (1998). Major ion chemistry ans control of weathering in a high altitude basin, Alakananda, Garhwal Himalaya, India. Hydrol. Sci. Jour., v.43, pp.825-844.

Singh, A.K. and Hasnain, S.I. (1999). Environmental geochemistry of Damodhar river basin, east coast of India. Environ. Geol., v.37(1-2), pp.124-136.

Walton, W.C. (1970) Groundwater resources evaluation. McGraw-Hill Book Co., New York.

WHO (1996) Guidelines for drinking water quality. v.2,Health criteria and other supporting information, WHO, Geneva. 973p.

Wilcox L.V (1955) Classification and use if irrigation waters, U.S.Department of Agriculture, Circulation, 969, Washington, D.C, pp.19.

Table 1: Chemical characteristics of groundwater in and around Salem magnesite mine area, Salem, Tamil Nadu, during summer (May) 1999.

S.No.	Well Type	Location	Depth (feet)	pH	EC (µS/cm)	TDS	Ca²⁺	Mg²⁺	Na⁺	K⁺	CO₃²⁻	HCO₃⁻	Cl⁻	SO₄²⁻	Fe	Mn	Ni	Cr	Co	%Na	SAR	RSC
									mg/L						µg/l					%		
2	Bore Well	Red Hills	200	9	696	650	44	45	24	0	54	262	10	27	23	2	39	192	34	15	0.6	0.2
4	Bore Well	Farm (Burn Std India Ltd)	150	8.6	889	1000	71	52	24	1	37	382	19	31	615	26	41	189	38	12	0.5	-0.4
6	Bore Well	Near Dalamia Factory	50	9.3	422	700	15	31	22	0	13	166	28	12	283	44	46	146	1	23	0.7	-0.1
7	Bore Well	Near Omalur High Road	120	8.4	976	1025	51	48	70	8	20	399	26	57	1336	36	45	95	1	33	1.7	0.7
8	Bore Well	Sambanayakkanpatti	250	8.8	1449	1975	58	75	120	8	46	574	62	55	590	60	50	89	1	37	2.4	1.9
9	Bore Well	Niranjipatti	250	9.6	1901	2000	82	71	202	9	74	658	172	11	2012	83	49	74	1	48	3.9	3.3
10	Bore Well	Mankuppai	200	8.1	1119	1350	27	53	116	16	24	420	78	13	497	34	45	78	1	49	3.0	2.0
11	Bore Well	Kambinayakkanpatti	200	7.8	1848	1700	78	94	143	23	37	701	145	14	1450	43	49	59	1	37	2.6	1.1
13	Bore Well	In between Palaiyanur & Kambinayakkanpatti	360	7.7	1317	900	70	66	87	17	37	481	76	40	3876	96	49	32	1	32	1.8	0.2
14	Bore Well	Kondamanayakkanpatti	320	7.6	1363	925	65	57	131	0	34	416	182	27	889	49	42	23	1	42	2.9	0.0
17	Bore Well	Kurumbapatti Children Park	200	8.4	924	1513	53	59	40	1	34	397	32	35	1615	76	46	47	1	42	2.9	0.9
19	Bore Well	Kobaipatti N.G.G.O Colony	200	7.2	879	950	56	46	51	1	8	228	70	123	1014	48	53	139	5	19	0.9	0.1
20	Bore Well	Shanti Nagar	200	7.1	795	675	53	39	48	0	20	263	70	48	317	20	54	99	6	25	1.2	-2.6
22	Bore Well	Kannankurichchi	50	7.2	1421	3500	28	88	126	3	23	171	122	328	295	137	61	169	16	26	1.2	-0.9
23	Bore Well	In between Kannankurichchi & Manakkadu	36	.7	1891	4100	26	95	217	14	9	419	156	306	1732	367	74	213	38	39	2.6	-5.1
24	Bore Well	Manakkadu	200	.7	1171	1275	57	42	122	3	12	262	131	142	494	32	51	163	15	52	4.4	-2.0
25	Bore Well	Opp. to Central Jail (Thirunagar)	200	7.1	1023	975	57	39	96	0	8	249	120	120	1306	129	63	40	21	46	3.0	-1.6
27	Bore Well	Alagapuram	150	7.1	1826	1625	65	41	266	3	8	514	303	32	1689	154	84	39	25	41	2.4	-1.7
29	Bore Well	Dalavaypatti	300	6.7	1831	2750	22	55	283	15	22	454	314	13	3026	241	80	59	43	64	6.4	2.1
30	Bore Well	Selattampatti	250	7	3090	5225	54	145	367	11	3	336	414	616	191	20	84	102	65	69	7.3	2.5
31	Bore Well	Vedakuttampatti	100	6.9	1095	1975	49	59	77	10	0	173	97	221	2401	245	61	45	29	53	5.9	-9.0
32	Bore Well	Andipatti	350	7	1238	1625	24	57	143	11	25	307	162	55	1785	84	64	40	21	33	1.8	-4.5
33	Bore Well	Mijyanteruvu	200	6.5	3154	4750	93	128	354	36	0	995	390	113	1032	128	88	97	54	52	3.6	0.0
34	Bore Well	Sivadapuram	150	6.4	2848	4525	31	146	337	11	85	198	373	532	823	198	75	76	40	52	5.6	1.1
35	Bore Well	Johnsonpet	200	6.5	1542	1850	37	66	186	3	18	220	165	296	853	70	74	43	24	52	5.6	-7.5
36	Bore Well	Ponnammapettai	200	6.6	1559	1900	28	45	237	6	4	455	163	137	467	28	73	32	14	53	4.2	-3.1
37	Bore Well	Ammapettai	150	6.6	1846	1900	37	53	259	40	0	411	238	143	3204	305	80	32	17	67	6.5	2.5
38	Bore Well	V.O.C. nagar	150	6.3	3141	3700	39	110	467	14	6	167	435	758	2599	157	102	90	34	67	6.4	0.6
39	Bore Well	Dr.K.K. Weavers Colony	200	7	1050	875	21	57	101	0	0	228	117	119	1735	89	67	16	32	65	8.7	-8.1
40	Bore Well	Erumapalayam	250	6.8	1441	1025	66	50	161	14	25	481	175	36	1007	56	61	48	16	45	2.6	-2.0
41	Bore Well	Katarampatti	150	6.4	2874	5500	42	96	429	4	10	373	314	624	3021	282	91	98	70	49	3.6	1.3
42	Bore Well	Silnayakkanpatti	200	6.3	735	800	30	40	59	0	11	185	104	49	3946	67	55	15	83	65	8.3	-3.6
43	Bore Well	Tatagapatti	200	6.3	1518	2400	48	37	222	3	2	312	219	167	387	38	76	55	50	35	1.7	-1.4
44	Bore Well	Gugai	200	6.8	1506	5375	55	52	180	7	8	460	298	39	1171	68	78	77	54	64	5.8	-0.2
45	Bore Well	Near Kondalampatti By pass Road	250	6.4	2938	2938	88	62	344	16	0	255	298	542	400	21	73	60	60	53	4.2	0.8
46	Bore Well	Annadanappatti	200	6.3	1546	1725	57	48	195	7	16	344	211	133	106	371	76	71	57	62	6.9	-5.3
47	Bore Well	Shevapettai	150	7	1394	1375	50	55	154	7	11	408	172	70	3784	242	58	29	31	56	4.6	-0.6
48	Bore Well	Arisipalayam	200	6.3	1682	1600	28	114	135	7	23	284	148	317	818	79	66	9	42	50	3.6	0.0
50	Bore Well	Omalur high road	200	8.8	1315	1325	58	72	98	1	28	488	105	50	486	26	58	12	32	36	2.5	-5.4
52	Bore Well	Salem Old Bus Stand	200	7.3	1335	350	73	24	136	70	0	353	150	9	210	20	18	61	25	33	2.0	0.1
55	Bore Well	Tyagampatti	250	5.7	1013	1800	27	67	74	1	8	313	85	104	1073	51	21	57	45	58	3.5	0.1
12	Dug Well	Palaiyanur	300	8.5	1664	1075	81	69	147	21	30	588	153	20	63	2	48	83	1	32	1.8	-1.5
15	Dug Well	Ashok Nagar (Paly Karadu)	70	8.5	1385	550	93	67	83	2	37	343	205	47	425	68	49	26	1	26	1.6	-3.3
16	Dug Well	Chettichawadi	15	8.7	1980	3000	39	46	284	67	35	873	11	47	257	19	61	103	1	71	7.3	9.8
18	Dug Well	Pallavapuram	200	7.4	1235	1275	58	36	148	2	45	387	69	109	110	13	58	128	3	53	3.8	2.0
21	Dug Well	Near Vercaud foot Hills	50	7.3	1079	850	53	41	109	0	25	318	101	91	73	22	52	113	10	44	2.7	0.0
26	Dug Well	Opp. to Saradha College	50	7.3	1592	1675	56	65	174	9	35	581	155	7	172	21	61	144	13	49	3.7	2.5
28	Dug Well	Opp. to Thyagaraja Polytechnic	50	7	1549	2300	33	62	196	9	6	334	286	50	73	14	23	70	37	56	4.6	-1.1
49	Dug Well	Swarnapuri Annexe	120	7.2	1482	1550	42	53	190	3	21	428	199	54	87	7	71	38	49	56	4.6	1.3
51	Dug Well	Ammapalayam	100	9	1784	1775	65	68	205	3	22	450	253	106	818	20	15	14	43	50	4.2	-0.7

The zero values indicates that the values are below the deduction limit of the instrument

Table 2: Chemical characteristics of groundwater in and around Salem magnesite mine area, Salem, Tamil Nadu, during winter (December) 1999

S.No	Well Type	Location	Depth (feet)	pH	EC (µS/cm)	TDS	Ca²⁺	Mg²⁺	Na⁺	K⁺	CO₃²⁻	HCO₃⁻	Cl⁻	SO₄²⁻	Fe	Mn	Ni	Cr	Co	%Na	SAR	RSC
									mg/L							µg/l				%		
2	Bore Well	Red Hills	200	n.a	0	0	0	0	0	0	0	0	0	0	0	0	1	0	0	0.0	0.0	0.0
4	Bore Well	Farm (Burn Std India Ltd)	150	8.1	923	1092	45	81	7	2	30	284	29	118	216	0	1	149	0	4	0.2	-3.2
6	Bore Well	Near Dalamia Factory	50	8.0	740	1092	54	50	14	1	19	217	26	108	327	34	12	106	7	8	0.3	-2.6
7	Bore Well	Near Omalur High Road	120	7.9	1054	1326	22	98	29	4	41	330	80	52	70	14	16	84	4	13	0.6	-2.4
8	Bore Well	Sambanayakkanpatti	250	n.a	0	0	0	0	0	0	0	0	0	0	0	0	0	0	0	0.0	0.0	0.0
9	Bore Well	Niranjipatti	250	8.0	2204	2106	45	195	84	5	8	700	276	96	2080	24	21	11	7	17	1.2	-6.5
10	Bore Well	Mankuppai	200	7.6	1490	1404	55	126	36	11	26	536	65	125	243	105	1	97	4	12	0.6	-3.4
11	Bore Well	Kambinayakkanpatti	200	7.5	2150	1950	55	191	59	18	59	579	214	139	728	15	1	35	2	14	0.8	-7.0
13	Bore Well	In between Palaiyanur & Kambinayakkanpatti	360	8.5	1449	1482	39	128	41	8	47	462	126	55	57	10	1	29	5	14	0.7	-3.3
14	Bore Well	Kondamannayakkanpatti	320	8.2	1010	1248	52	65	49	1	26	357	50	83	1989	93	34	24	29	21	1.1	-1.2
17	Bore Well	Kurumbapatti Children Park	200	8.2	967	1170	22	96	15	1	11	353	38	110	372	16	1	81	1	7	0.3	-2.9
19	Bore Well	Kobaipatti N.G.G.O Colony	200	7.6	988	1170	95	55	34	1	0	273	62	167	3451	54	37	103	19	6	0.3	-4.8
20	Bore Well	Shanti Nagar	200	7.4	921	1248	61	59	30	1	26	257	66	98	1251	21	9	97	5	15	0.7	-2.8
22	Bore Well	Kannankurichchi	50	7.7	2780	2340	50	278	55	2	50	336	562	216	96	16	32	123	7	9	0.7	-18.2
23	Bore Well	In between Kannankurichchi & Manakkadu	36	7.5	2447	2340	152	162	68	24	22	262	620	27	136	29	37	153	18	15	0.9	-15.9
24	Bore Well	Manakkadu	200	7.4	1393	1716	99	74	64	3	9	242	225	142	345	20	19	113	8	21	1.2	-6.8
25	Bore Well	Opp. to Central Jail (Th runagar)	200	7.3	1019	1326	98	44	38	1	23	257	86	125	291	18	22	97	2	16	0.8	-3.6
27	Bore Well	Alagapuram	150	7.3	1908	1872	119	111	91	1	50	472	325	11	307	25	20	22	9	21	1.4	-5.7
29	Bore Well	Dalavaypatti	300	7.2	2862	2496	163	190	106	12	26	647	568	12	313	275	37	62	26	17	1.3	-12.3
30	Bore Well	Selattampatti	250	7.6	4261	3900	128	367	133	9	19	289	1218	105	86	28	42	103	41	14	1.4	-31.2
31	Bore Well	Vedakuttampatti	100	7.8	1820	1716	130	112	53	5	8	253	292	244	126	25	46	63	27	13	1.4	-11.3
32	Bore Well	Andipatti	350	7.6	1961	1638	53	176	53	6	28	280	259	301	693	25	11	26	12	13	0.8	-11.6
33	Bore Well	Miyanteruvu	200	7.3	3542	3432	165	266	112	17	16	348	990	7	479	38	37	93	18	15	1.3	-23.9
34	Bore Well	Sivadapuram	150	7.5	3191	2418	118	276	74	5	52	520	660	121	496	127	21	74	1	11	0.9	-18.3
35	Bore Well	Johnsonpet	200	7.2	1268	1716	98	59	67	2	0	228	253	73	1235	27	40	32	1	23	1.3	-6.0
36	Bore Well	Ponnammapettai	200	7.2	1701	1950	96	105	80	4	25	406	268	74	898	33	34	30	1	21	1.3	-6.0
37	Bore Well	Ammapettai	150	7.1	2823	3432	63	261	78	9	43	313	298	604	343	24	58	23	2	13	1.0	-18.1
38	Bore Well	V.O.C. nagar	150	7.0	3142	3432	146	226	126	3	25	555	704	61	2537	71	44	41	1	18	1.5	-15.9
39	Bore Well	Dr.K.K. Weavers Colony	200	7.3	1239	1560	71	77	59	1	19	269	222	41	3640	51	9	58	1	21	1.2	-4.8
40	Bore Well	Erumapalayam	250	7.1	1299	1482	106	61	63	1	19	385	154	71	911	33	30	40	1	21	1.2	-3.3
41	Bore Well	Katarampatti	150	7.3	3019	3510	127	231	110	3	45	431	637	156	1303	35	74	52	19	16	1.3	-16.8
42	Bore Well	Silnayakkanpatti	200	7.1	798	1092	72	44	41	2	50	322	153	181	3875	293	41	61	10	10	0.4	-5.4
43	Bore Well	Tatagapatti	200	7.0	2274	2574	165	124	97	3	0	359	545	54	574	129	41	85	17	19	1.4	-12.6
44	Bore Well	Gugai	200	7.3	2686	2574	154	179	97	8	31	394	626	50	1543	190	38	111	24	17	1.3	-14.9
45	Bore Well	Near Kondalampatti By pass Road	250	7.1	3272	3042	236	202	97	7	17	330	888	60	167	33	51	86	21	13	1.1	-22.4
46	Bore Well	Annadanappatti	200	7.3	1441	1716	89	85	65	4	41	270	240	69	1024	31	27	44	1	20	1.2	-5.7
47	Bore Well	Shevapettai	150	7.2	1500	1716	91	97	54	3	9	318	241	112	410	64	31	12	1	16	0.9	-7.0
48	Bore Well	Arisipalayam	200	7.2	1484	1638	83	104	48	2	0	300	246	132	350	22	51	26	1	14	0.8	-7.8
50	Bore Well	Omalur high road	200	7.6	1571	1560	63	131	41	2	50	387	166	132	856	29	31	61	1	12	0.7	-5.9
52	Bore Well	Salem Old Bus Stand	200	7.5	1379	1638	93	67	64	32	31	322	153	68	218	22	18	10	1	26	1.2	-3.9
55	Bore Well	Tyagampatti	250	7.3	1624	1560	122	104	37	1	23	389	154	219	2740	53	43	11	3	10	0.6	-7.5
12	Dug Well	Palaiyanur	300	8.0	2062	1794	87	156	72	11	82	436	317	49	49	9	25	79	20	17	1.1	-7.3
15	Dug Well	Ashok Nagar (Paly Karadu)	70	7.8	991	1170	70	63	28	1	43	325	42	84	209	21	21	44	7	12	0.6	-1.9
16	Dug Well	Chettichawadi	15	8.0	2050	2496	85	124	108	56	84	174	417	20	425	15	22	85	20	30	1.7	-8.8
18	Dug Well	Pallavapuram	200	7.6	1131	1326	58	85	32	1	35	365	54	116	82	15	17	91	10	13	0.6	-2.8
21	Dug Well	Near Yercaud foot Hills	50	7.4	813	1248	45	48	44	1	26	309	46	33	97	24	4	63	2	24	1.1	-0.3
26	Dug Well	Opp. to Saradha College	50	7.8	1847	1794	64	148	66	9	50	489	230	75	316	19	5	35	14	17	1.0	-5.7
28	Dug Well	Opp. to Thyagaraja Polytechnic	50	7.8	3143	2652	176	228	85	8	11	622	682	46	300	28	30	67	33	12	1.0	-17.0
49	Dug Well	Swarnapuri Annexe	120	8.0	2012	1872	83	156	71	3	41	458	297	119	108	29	40	52	5	16	1.1	-8.1
51	Dug Well	Ammapalayam	100	7.3	1321	1638	24	118	52	2	38	312	206	36	3774	70	19	24	1	18	1.0	-4.5

For Sl. No. 2 and 8 zero (0) denotes that samples have not been collected. Zero (0) at other places stands for the values below deduction limit of the instrument.

Table 3: Range in values of chemical parameters in the study area and WHO and Indian Standards for drinking water

Parameters	Range in the study area in summer		Range in the study area in summer		Maximum permissible limit(WHO & ISI)
	Tube well	Dug well	Tube well	Dug well	
pH	5.7-9.6	7.0-8.7	7-8.5	7.3-8.0	6.5 to 8.5
Ec	422-3154	1079-1980	740-4261	813-4143	
TDS	350-5500	550-3000	1092-3900	1170-2652	500.0
Ca	15-93	33-93	22-237	45-176	75.0
Mg	24-146	41-69	44-367	48-228	30.0
Na	22-467	83-284	7-133	28-108	<200
K	0-70	0-67	1-24	1-56	<10
HCO3-	166-995	334-873	109-700	174-622	<500
SO42-	9-758	7-109	7-604	20-119	150.0
CO32-	0-85	16 - 45	0-59	11-84	<10
Cl-	10-435	11-286	26-1218	42-682	250.0
Fe	106-3946	63-818	57-3875	49-3774	0.3
Mn	2-367	2-68	4-293	9-70	0.1
Ni	18-102	15-71	1-74	4-40	0.1
Cr	9-213	14-144	11-153	24-91	0.1
Co	1-83	1-49	1-41	1-33	0.1
%Na	12-69	26-71	4-26	12-30	
SAR	0.5-8.7	1.6-7.3	0.2-1.5	0.6-1.7	
RSC	(-0.9)-(3.3)	(-3.3)-(9.8)	(-31.2)-(-2.4)	(-17.0)-(-0.3)	

Units: Conc. In mg/l except pH; EC (µS/cm); Fe,Mn,Ni,Cr,Co in mg/l; RSC and SAR in meq/l

Table 4: Characterisation of groundwater on the basis of Piper Trilinear Diagram (Piper, 1944)

Sub-division of the diamond shaped field	Charateristics of corresponding sub-divisions of diamond shaped fields
1	Alkaline earths (Ca+Mg) exceeds alkalies (Na+K)
2	Alkalies exceeds alkaline earths
3	Weak acids (CO3+HCO3) exceeds strong acids (SO4+Cl+F)
4	Strong acids exceed weak acids
5	ness (secondary alkalinity) exceeds 50% (Chemical pro... hs and weak acids
6	Non-carbonate hardness (secondary salinity) exceeds 50% (Chemical properties are dominated by alkaline earths and strong acids)
7	Carbonate alkali (primary salinity) exceeds 50% (Chemical properties are dominated by alkalies and weak acids)
8	Carbonate alkali (primary alkalinity) exceeds 50% (Chemical properties are dominated by alkalies and weak acids)
9	No one cation-anion pair exceeds 50%

210

Intl. Conf. on Sustainable Development and Management of Groundwater Resources
in Semi-Arid Region with Special Reference to Hard Rock, (IGC 2002),
M. Thangarajan, S.N. Rai & V.S. Singh (Eds.)

Parametric studies on the effect of field parameters on seawater intrusion in multi-layered coastal aquifers

A. K. Rastogi[1] and S.K. Ukarande[2]

[1]Department of Civil Engineering, Indian Institute of Technology, Powai, Mumbai-400076, India
e-mail: akr@civil.iitb.ac.in

[2]Department of Civil Engineering, MGM's College of Engineering & Technology, Kamothe,
Navi Mumbai-410209, India
e-mail: ukarande@yahoo.com

Abstract

A numerical model is developed to study two-dimensional steady state seawater intrusion problem involving hydrodynamic dispersion in a multi-layered confined coastal aquifer. In this approach two non-linear partial differential equations of flow and solute transport are coupled. These equations are expressed in terms of freshwater head and solute concentration where, both, are dependent variables. Solutions are obtained by Galerkin's finite element method. The results are compared with some of the existing solutions. The verified model is extended to determine the influence of various field parameters, namely, discharge, dispersivity, hydraulic conductivity on seawater intrusion for single and multilayered coastal aquifer.

Key words: Numerical model, coastal aquifer, seawater intrusion, dispersion

Introduction

Coastal aquifers usually suffer from the danger of seawater intrusion and consequent deterioration of quality of its freshwater. India does possesses a very long coastline and all along this it has witnessed a very rapid growth involving, both, urbanization and industrial development. This has resulted in sudden spurt in the population in these sectors causing increased freshwater demand that requires design of new ground water supply schemes in the coastal regions. This largely forces increased withdrawal of ground water, which opens up the possibility of contamination of its quality due to salt water intrusion. Proper planning of withdrawal schemes as well as replenishment of these aquifers with suitable recharging methods is an imperative requirement in the coastal regions.

Decline of water table consequent to the overstraining of coastal groundwater resources has been attributed to be the major cause of seawater intrusion into coastal aquifers. In India the problem of coastal seawater intrusion is of serious concern in the littoral states of Gujarat, Kerala, Tamil Nadu and West Bengal. In view of the long Indian coastline, seawater intrusion control needs greater emphasis in water resources management.

In order to assess the contamination and plan the remedial measures, it is necessary to determine the extent to which the ingression of seawater occurs and the resulting concentration distribution of salt in the coastal aquifer subject to the imposed conditions. A flexible mathematical model capable of predicting saltwater concentration in the aquifer over a wide range of field conditions constitutes an important tool in the groundwater management program of coastal aquifers.

Many intrusion models are based on Ghyben-Herzberg relationship assuming freshwater and seawater are immiscible and has a well defined interface. Since the two fluids are miscible, a sharp interface concept is not realistic especially when thickness of dispersion zone is considerable (Bear, 1979). Presently a synthetic rectangular confined aquifer is considered involving a set of realistic boundary conditions. The simulated steady state solutions are compared with (Henry, 1964), (Pinder and Cooper, 1970) and (Lee and Cheng, 1974) to assess the correctness of the developed algorithm. This is further modified to examine the influence of multilayer coastal aquifer and the spread of diffused interface. The model is used to examine the influence of various field parameters on seawater intrusion for the single and multilayered confined coastal aquifer.

Development of a Numerical Model

Governing Equations

The equations describing steady flow and solute transport in an isotropic porous medium can be written as:

Darcy's Equation:

$$q_i = \frac{-k_{ij}}{\mu}\left[\frac{\partial p}{\partial x_j} + \rho.g.e_j\right] \qquad (1)$$

Flow Equation:

$$\frac{\partial}{\partial x_i}(\rho q_i) = 0 \qquad (2)$$

Solute Transport Equation:

$$\frac{\partial}{\partial x_i}\left(D_{ij}\frac{\partial C_i}{\partial x_j}\right) - V_i\frac{\partial C}{\partial x_i} = 0 \qquad (3)$$

The constitutive equation relating fluid density to concentration is:

$$\rho = \rho_f(1 + \varepsilon C) \qquad (4)$$

Here, V_i are the components of the seepage velocity [LT^{-1}], k_{ij} is the intrinsic permeability of the porous medium [L^2], μ is the dynamic viscosity of the fluid [$ML^{-1}T^{-1}$], ρ is the fluid density [ML^{-3}], p is the fluid pressure [$ML^{-1}T^{-2}$], e_j are the components of the gravitational unit vector [LT^{-2}], D_{ij} is the dispersion coefficient [$L^2 T^{-1}$] and C is the concentration of the pollutant [ML^{-3}] and C is the density difference ratio.

All the symbols used in this paper are defined in the list of nomenclature.

On substituting the following

$$h = \frac{p}{\rho_f g} + z \;\; ; \;\; \varepsilon = \frac{\rho_s - \rho_f}{\rho_f} \;\; ;$$

$$\frac{\rho}{\rho_f} - 1 = \rho_r = \varepsilon C \qquad (5)$$

in equation (1) we get,

$$q_i = -K_{ij}\left[\frac{\partial h}{\partial x_j} + \varepsilon C \; e_j\right] \qquad (6)$$

where ρ_f is the freshwater density, g is the acceleration due to gravity and z is the elevation above datum.

Introducing the non-dimensional variables,

$$x' = \frac{x}{d} \;\; ; \qquad z' = \frac{z}{d} \;\; ; \qquad h' = \frac{h}{d} \;\; ;$$

$$C' = \frac{C}{C^*} \;\; ; \;\; K'_{Lzz} = \frac{K_{zz}}{K_{Lxx}} \;\; ; \;\; K'_{Lxx} = \frac{K_{xx}}{K_{Lxx}} \;\; ;$$

$$V'_x = \frac{V_x}{V} \;\; ; \;\; V'_z = \frac{V_z}{V} \;\; ; \qquad \alpha'_T = \frac{\alpha_T}{\alpha_L} \;\; ;$$

$$\alpha'_L = \frac{\alpha_L}{\alpha_L} = 1 \quad and \quad d_L = \frac{d}{\alpha_L}$$

in equation (2) and (3) and dropping the primes on all the terms in further analysis we get,

$$K_{Lxx}\frac{\partial^2 h}{\partial x^2} + K_{Lzz}\frac{\partial^2 h}{\partial z^2} = -K_{Lzz}\varepsilon\frac{\partial C}{\partial z} \qquad (7)$$

$$D_{xx}\frac{\partial^2 C}{\partial x^2} + D_{zz}\frac{\partial^2 C}{\partial z^2} + D_{xz}\frac{\partial^2 C}{\partial x\partial z}$$

$$ \qquad (8)$$

$$+ D_{zx}\frac{\partial^2 C}{\partial z\partial x} - \left(V_x\frac{\partial C}{\partial x} + V_z\frac{\partial C}{\partial z}\right)d_L = 0$$

where d is the depth of confined coastal aquifer, C^* is concentration of seawater, α_L and α_T are longitudinal and transverse (vertical) dispersivities of the medium respectively, V is the average linear velocity, K_{Lxx} and K_{Lzz} are non-dimensional hydraulic conductivities along x and z direction respectively, and, D_{ij} is the dispersion tensor, whose terms in non-dimensional form are defined in two-dimensional co-ordinate (x-z) system as:

$$D_{xx} = \frac{V_x^2}{|V|} + \alpha_T \frac{V_z^2}{|V|}$$

$$D_{zz} = \frac{V_z^2}{|V|} + \alpha_T \frac{V_x^2}{|V|} \qquad (9)$$

$$D_{xz} = D_{zx} = (1 - \alpha_T)\frac{V_x V_z}{|V|}$$

The average linear velocity (i.e. V_x and V_z) in non-dimensional form for 2-D x-z coordinate system can be written as;

$$\therefore V_x = -\frac{A}{\theta} K_{Lxx}\left(\frac{\partial h}{\partial x}\right) \qquad (10)$$

$$\therefore V_z = -\frac{A}{\theta} K_{Lzz}\left(\frac{\partial h}{\partial z} + \varepsilon C\right) \qquad (11)$$

where $V = \dfrac{K_{Lii}}{A}$, and, $\dfrac{1}{A}$ is piezometric gradient.

Thus the equations (7), (8), (10) and (11) are Governing Equations in non-dimensional form.

Solution Domain, Initial and Boundary Conditions

Solution Domain

The rectangular two-dimensional vertical cross-section of length L and depth d of the synthetic confined coastal aquifer involves two strata. The two aquifer layers are speared by a semi pervious layer which extends from x/L = 0.0 to 2.0 between z/d = 0.25 to 0.5, in which the flow is considered to be vertical as shown in fig.1.

Initial Conditions

1. C = 0.0 on the upstream boundary where freshwater is flowing into the aquifer.

2. C = 1.0 on seaward boundary.

3. Initially the aquifer is assumed to contain freshwater (i.e. C = 0.0) everywhere in the aquifer.

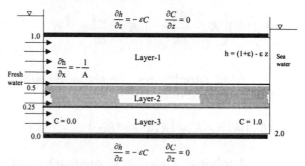

Figure 1 Definition sketch and boundary conditions

Boundary Conditions

Boundary conditions are shown in fig.1.

Finite Element Formulation

The solution to the flow and solute transport equations with appropriate boundary conditions is obtained by the finite element method using Galerkin's approach. Triangular elements are used with linear interpolation functions, which define the variation of hydraulic head and salt concentration within an element in terms of their nodal values. Presently the entire flow region is divided in 256 elements and 153 nodes.

In case of linear triangular element the field variables h and C are given as:

$$h(x,z) = N_i h_i + N_j h_j + N_k h_k$$
$$C(x,z) = N_i C_i + N_j C_j + N_k C_k \qquad (12)$$

where N_i, N_j and N_k are the interpolation functions for the three nodes of an element.

Applying Galerkin's approach and Green's Theorem, equation (7) becomes,

$$\int_A K_{Lxx} \frac{\partial\{N\}}{\partial x} \frac{\partial[N]}{\partial x} \{h\} \ dA$$

$$+ K_{Lzz} \int_A \frac{\partial\{N\}}{\partial z} \frac{\partial[N]}{\partial z} \{h\} \ dA$$

$$= \varepsilon.K_{Lzz} \int_A \{N\} \frac{\partial[N]}{\partial z} \{C\} dA \qquad (13)$$

$$+ \int_S \{N\} \left(K_{Lxx} \frac{\partial h}{\partial x} n_x + K_{Lzz} \frac{\partial h}{\partial z} n_z \right) ds$$

After simplifying, the above equation can be written in matrix form as

$$[K_3]\{h\} = \varepsilon.K_{Lzz}[R]\{C\} + \{Bh_3\} \qquad (14)$$

Applying Galerkin's approach and Green's Theorem, equation (8) becomes,

$$\int_A \left(D_{xx} \frac{\partial\{N\}}{\partial x} . \frac{\partial[N]}{\partial x} . \{C\} + D_{zz} \frac{\partial\{N\}}{\partial z} . \frac{\partial[N]}{\partial z} . \{C\} \right.$$

$$+ D_{xz} \frac{\partial\{N\}}{\partial x} . \frac{\partial[N]}{\partial z} . \{C\} + D_{zx} \frac{\partial\{N\}}{\partial z} . \frac{\partial[N]}{\partial x} . \{C\}$$

$$\left. + d_L \{N\} V_x \frac{\partial[N]}{\partial x} \{C\} + d_L \{N\} V_z \frac{\partial[N]}{\partial z} \{C\} \right) dA$$

$$= \int_S (D_{xx} \frac{\partial C}{\partial x} n_x + D_{zz} \frac{\partial C}{\partial z} n_z + D_{xz} \frac{\partial C}{\partial x} n_x$$

$$+ D_{zx} \frac{\partial C}{\partial x} n_z) \ \{N\} \ ds \qquad (15)$$

After simplification the above equation can be written in matrix form as,

$$[K_4] \ \{C\} = \{B_c\} \qquad (16)$$

Discretizing equation (10) and applying Galerkin's approach, we get

$$\int_A \{N\} \left[V_x + \frac{A}{\theta} K_{Lxx} \frac{\partial h}{\partial x} \right] dA = 0 \qquad (17)$$

The above equation can be written in matrix form as,

$$[M]\{V_x\} = \{F_1\} \qquad (18)$$

Similarly discretizing equation (11) and applying Galerkin's approach, we get

$$\int_A \{N\} \left[V_z + \frac{A}{\theta} K_{Lzz} \left(\frac{\partial h}{\partial z} + \varepsilon C \right) \right] dA = 0$$

$$(19)$$

The above equation can be written in matrix form as,

$$[M]\{V_z\} = \{E_1\} \qquad (20)$$

In eqns. (14, (16), (19) and (20) $[K_3]$, $[K_4]$ and $[M]$ are coefficient matrices and, $[R]$ $\{C\}$, $\{Bh_3\}$, $\{Bc\}$, $\{F_1\}$, $\{E_1\}$ are column vectors containing boundary conditions.

Values of Field Parameters

The seawater intrusion including velocity dependent dispersion depends upon four non-dimensional parameters namely α_T, K (K_{Lxx}, K_{Lzz}), d_L and A which are included in the governing equations. For many field problems of practical interest the transverse (vertical) hydraulic conductivity is found to be less than the longitudinal hydraulic conductivity. In view of this consideration the range for non-dimensional K (K_{Lxx} and K_{Lzz}) is considered presently to vary from 0.1 to an upper limit of 1.0. Parameter (A) is considered as inverse of upstream hydraulic gradient

$$\left(i.e. \ \frac{1}{A} = \frac{\partial h}{\partial x} \right).$$

The slope of the piezometric surface (hydraulic gradient) varies from 0.01 % to 0.5% for realistic field situations. This establishes upper and lower bounds for parameter A as 10,000 to 200 respectively. In view of the fact that depth d of the many coastal aquifers vary from 1 to 500 m, a range of 0.01 to 20 is assigned for the value of d_L, keeping in mind that larger aquifers should also have larger dispersivities (Pandit and Anand, 1984). Transverse dispersivity is normally an order of magnitude less than the longitudinal dispersivity. Therefore bounds of 0.1 to an upper limit of 1.0 for non-dimensional parameter ($\propto_T$) are considered reasonable for the present problem.

Results and Discussion

Equation (14) is solved for the nodal values of hydraulic head, assuming initial salt concentration (C = 0.0) distribution in the aquifer. The velocity vector distribution can be obtained by equations (18) and (20) utilizing the head values obtained earlier. Following this equation (16) is solved for nodal concentration values using the velocity vectors and dispersion coefficients defined by equation (9). The concentration values obtained are resubstituted in the equation (14) and this iterative cycle is repeated until a predetermined level of accuracy (specified tolerance limit = 0.01) is attained. Gaussian elimination method is used to solve the equations simultaneously. Solution to the seawater intrusion problem is obtained by Head Concentration (h-c) formulation involving the velocity dependent dispersion process. In further discussion, seawater wedge is chosen to be represented by a (C/C* = 0.5) isochlor, which is an approach adopted by Pandit and Anand (1984).

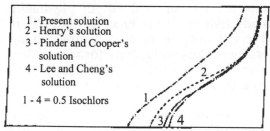

Figure 2: Comparison of positions of different 0.5 isochlors

First, for the homogenous isotropic single layer confined aquifer the results are compared with some known solutions. Fig.2 illustrates the comparison of positions of 0.5 isochlor obtained by the present model with the solutions of Henry (1964), Lee and Cheng (1974) and Pinder

and Cooper (1970). The present model results match closely with the existing solutions. However, the position of 0.5 isochlor is shifted towards landward side as compared to the existing solutions though the shape of the interface is S type and similar to the known shapes. This can be attributed to indicate that intrusion at the bottom of aquifer is more and relatively less at the top. The shift of seawater wedge towards landward side may be because of velocity dependent dispersion effects whereas in the previous models the dispersion coefficient is considered a fixed value independent of the velocity.

Parametric Study - Effect of Various Field Parameters on Seawater Intrusion

The finite element formulation developed involved four non-dimensional field parameters whose magnitude determines the extent of seawater ingress in the multilayered confined coastal aquifer. The main aim of this simulation is to understand qualitatively the behavior of seawater intrusion in multilayered confined coastal aquifer.

To examine the spread of 0.5 isochlor interface in the multilayer coastal aquifer, the flow region is divided in three layers, where the low hydraulic conductivity layer is sandwiched between two highly permeable aquifer zones. Fig.1 shows the definition sketch and boundary conditions for this case.

Effect of K_{Lxx} and K_{Lzz} on Seawater Intrusion

Table 1: Parametric data considered to study effect of K_{Lxx} and K_{Lzz} on seawater intrusion.

Set No.	$K_{Lx1} = K_{Lx3}$	K_{Lx2}	$K_{Lz1} = K_{Lz3}$	K_{Lz2}
1	1.0	1.0	1.0	1.0
2	1.0	0.01	1.0	1.0
3	1.0	0.01	1.0	0.01
4	1.0	0.1	1.0	0.1
5	1.0	0.05	0.5	0.05
6	1.0	0.01	0.1	0.01

The effect of non-dimensional values of K_{Lxx} and K_{Lzz} is illustrated in fig. 3 by keeping other parameters (A = 200, d_L = 2.0 α_T = 0.1 and θ = 0.4) constant for the multilayered coastal aquifer. For studying the effect of K_{Lxx} and K_{Lzz} on seawater intrusion, the sets of values considered are as given in Table-1.

The non-dimensional parameters K_{Lxx} and K_{Lzz} for aquifer layer-1, 2 and 3 are explained as follows:

$$K_{lxx} = \frac{K_{lx1}}{K_{lx1}} \quad \dots \text{(for layer}-1\text{)}, \quad \text{say } K_{lx1}$$

$$K_{lzz} = \frac{K_{lz1}}{K_{lx1}} \quad \dots \text{(for layer}-1\text{)}, \quad \text{say } K_{lz1}$$

$$K_{lxx} = \frac{K_{lx2}}{K_{lx1}} \quad \dots \text{(for layer}-2\text{)}, \quad \text{say } K_{lx2}$$

$$K_{lzz} = \frac{K_{lz2}}{K_{lx1}} \quad \dots \text{(for layer}-2\text{)}, \text{say } K_{lz2}$$

$$K_{lxx} = \frac{K_{lx3}}{K_{lx1}} \quad \dots \text{(for layer}-3\text{)}, \quad \text{say } K_{lx3}$$

$$K_{lzz} = \frac{K_{lz3}}{K_{lx1}} \quad \dots \text{(for layer}-3\text{)}, \quad \text{say } K_{lz3}$$

where, K_{Lxx} and K_{Lzz} are hydraulic conductivities in non-dimensional form along x and z directions, respectively. In the further study K_{Lx1}, K_{Lx2}, K_{Lx3}, K_{Lz1}, K_{Lz2} and K_{Lz3} are used as notations for non-dimensional K_{Lxx} and K_{Lzz} parameters for convenience for respective layers. Subscripts 1, 2 and 3 refer to layer-1, layer-2 and layer-3 respectively. In this analysis K_{Lx1} is equal to K_{Lx3} and K_{Lz1} is equal to K_{Lz3}.

Set-1 values for hydraulic conductivities indicate that the aquifer is single layered and effect of the same is illustrated in fig.3 by keeping other field parameters (A = 200, d_L = 1.0, α_T = 0.1 and θ = 0.4) constant. In Set-2, the value of K_{Lx2} is being reduced to 0.01 by keeping other values same, which indicates flow can be vertical in the layer-2. Set-3 and Set-4 show reduction in values for the hydraulic conductivities for layer-2 (K_{Lx2} and K_{Lz2}) to 1/100 and 1/10, respectively, as compared to that of layer-1 and layer-3. It indicates that the hydraulic conductivity for layer-2 is less as compared to the upper and lower layers. The presence of layer-2 involving less hydraulic conductivity leads to greater intrusion of seawater into coastal aquifers as it obstructs free movement of flow from layer-1 to layer-3 and vice versa. Further decrease in values of K_{Lxx} and K_{Lzz} for layer-1 and layer-3 along with layer-2, leads to more intrusion of seawater into coastal aquifers.

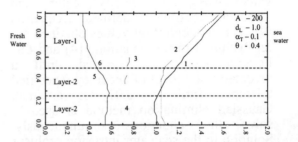

Figure 3: Effect of parameters K_{Lxx} and K_{Lzz} on seawater intrusion for the parametric data as shown in Table −1 for various sets of values (0.5 isochlor)

Above study shows the influence of various values of K_{Lxx} and K_{Lzz} on seawater intrusion and it helps to decide a suitable set of values of hydraulic conductivities for layered aquifer. It is observed that for layered aquifer having less hydraulic conductivity for sandwiched layer, of the same aquifer domain the seawater intrudes more in all the layers.

Above study therefore helps the author to consider Set-1, Set-2 and Set-4 values for K_{Lxx} and K_{Lzz} to determine the influence of other field parameters which are referred as Set-A, Set-B and Set-C in further analysis

Effect of A on Seawater Intrusion

The effect of A on seawater intrusion is illustrated in fig. 4-a-c, for three sets of specific values of field parameters, while A varies from 200 to 10,000. The three sets of values are shown in Table – 2

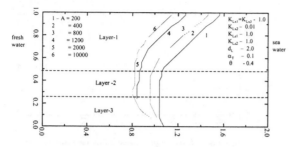

Figure 4 - b. Effect of parameter A on seawater intrusion (0.5 isochlor)

Table 2: Parametric data considered to study the effect of 'A' while it varies from 200 to 10,000 for $\alpha_T = 0.1$, $d_L = 2.0$ and $\theta = 0.4$ as constant values.

	K_{Lx1}	K_{Lx2}	K_{Lz1}	K_{Lz2}
Set-A	1.0	1.0	1.0	1.0
Set-B	1.0	0.01	1.0	1.0
Set-C	1.0	0.1	1.0	0.1

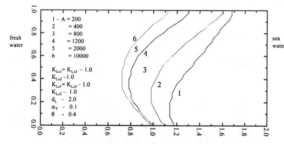

Figure – 4-a. Effect of parameter 'A' on seawater intrusion (0.5 isochlor)

The upstream hydraulic gradient is defined as the inverse of parameter A. Increase in the value of 'A' means the decrease in upstream head gradient which corresponds to a decrease in the freshwater flow to aquifer since the flow area and hydraulic conductivity remains unchanged and vice versa. The effect of 'A' on intrusion is illustrated in fig. 4-a for Set-A i.e. for single layer confined aquifer, while A varies from 200 to 10,000. The seawater intrusion increases with the increase in the value of A. Decrease in the upstream hydraulic gradient indicates decrease in

freshwater discharge upstream, leading to the movement of 0.5 isochors landward. The present study, therefore, indicates that for larger values of A, the seawater further intrudes in the coastal aquifer.

It is observed from the fig. 4 -b that with the increase in value of A from 200 to 10,000 the seawater intrusion increases for Set-B. Decrease in freshwater flow leads to the movement of 0.5 isochlor land ward. Study therefore finds that for larger values of A, the seawater intrudes further in the coastal aquifer for Set-B.

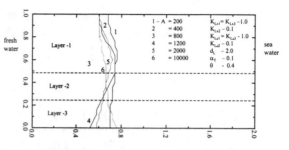

Figure 4 -c. Effect of parameter A on seawater intrusion (0.5 isochlor)

It is also observed from the fig. 4c that for Set-C, seawater intrusion increases in all the three layers and rate of intrusion is uniform. This may be due to less hydraulic conductivity for layer-2. But, this increased intrusion does not vary much with increase in value of A from 200 to 10,000. It shows that seawater intrudes more for Set-C (as compared to Set-B), but it does not have significant influence of larger values of A.

Effect of α_T on Seawater Intrusion

The influence of α_T which varies from 0.1 to 1.0 on seawater intrusion is determined and is shown in fig. 5 a-c keeping other field parameters constant. The three sets of values are as given in Table 3.

The effect of α_T on seawater intrusion is shown in fig.5-a for Set-A values while α_T varied from 0.1 to 1.0. It is observed from the figure that for the increase in value of α_T the seawater intrudes further into the aquifer. The non-dimensional parameter α_T is the ratio of the dimensional transverse dispersivity (α_T) to the longitudinal dispersivity (α_L). For small value of non-dimensional α_T, corresponds to larger coefficient of dispersivity in the longitudinal direction than in the transverse direction. Thus salt concentration spreads faster in the longitudinal direction, making the 0.5 isochlor more S-shaped, resulting in less intrusion near the top and more at the bottom of the aquifer. In all field problems longitudinal dispersivity is greater than transverse and the α_T may more realistically vary between a range from 0.1 to 0.33.

Table 3: Parametric data considered to study the effect of ' α_T ' while it varies from 0.1 to 1.0 for A = 200, d_L = 2.0 and θ = 0.4 as constant values.

	K_{Lx1}	K_{Lx2}	K_{Lz1}	K_{Lz2}
Set-A	1.0	1.0	1.0	1.0
Set-B	1.0	0.01	1.0	1.0
Set-C	1.0	0.1	1.0	0.1

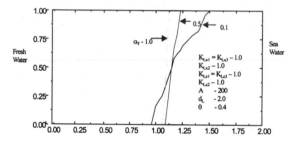

Figure 5-a Effect of parameter α_T on seawater intrusion (0.5-isochlor)

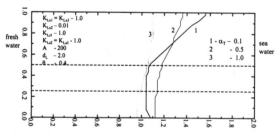

Figure 5-b . Effect of parameter α_T on seawater intrusion (0.5 isochlor)

The behavior of shape of 0.5 isochlor as explained above suits closely for Set-B values too (fig.5-b). For the Set-C values seawater intrudes more in all the layers (refer fig. 5-c) as layer-2 interferes

the cyclic flow and there is no significant influence on the shape and position of 0.5 isochlor while α_T varies from 0.1 to 1.0.

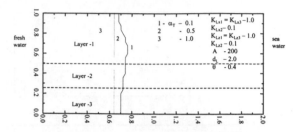

Figure 5 - c Effect of parameter α_T on seawater intrusion (0.5 isochlor)

Effect of d_L on Seawater Intrusion

Fig. 6 a-c illustrates the effect of d_L which varies from 0.1 to 8.0 on seawater intrusion for three sets of field parameters as shown in Table 4

Table 4 Parametric data considered to study the effect of 'd_L' while it varies from 0.1 to 8.0 for A = 200, α_T = 0.1 and θ = 0.4 as constant values.

	K_{Lx1}	K_{Lx2}	K_{Lz1}	K_{Lz2}
Set-A	1.0	1.0	1.0	1.0
Set-B	1.0	0.01	1.0	1.0
Set-C	1.0	0.1	1.0	0.1

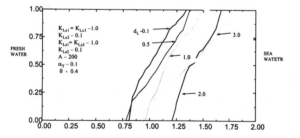

Figure 6 -a Effect of parameter d_L on seawater intrusion (0.5 - isochlor)

It is observed from the fig. 6-a, for Set-A values that as d_L increases from 0.1 to 1.0, the shape of 0.5 isochlor almost remains same, but its location shifts towards the seaward side. For the d_L equal to 2.0 and higher, 0.5 isochlor begins to take S – shape, i.e. intrusion of seawater increases at the bottom of the aquifer and reduces at the top.

218

Similarly it is observed from the fig. 6-b for Set-B values that as d_L varies from 0.1 to 4.0, the shape of 0.5 isochlor remains the same with considerable reduction in intrusion. For the d_L equal to 5.0 and higher, 0.5 isochlor begins to take S – shape, i.e. intrusion of seawater increases at the bottom and reduces at the top of the domain.

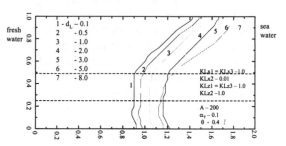

Figure 6-b . Effect of parameter d_L on seawater intrusion (0.5 isochlor)

The hydrodynamic dispersion coefficient depends on longitudinal and transverse dispersivities and pore velocities. The value of d_L increases presently with a decrease in the coefficient of longitudinal dispersivity (α_L). An increase in value of d_L from 0.1 to 4.0 corresponds to decreasing of α_L, since aquifer depth d is invariant value. Consequently hydrodynamic dispersion coefficient also decreases leading to lesser intrusion in the aquifer. However for further increase in value of d_L to 5.0 or more dispersion effects due to reduction in α_L becomes negligible with respect to convective effects and the hydraulic effects governing the cyclic flow become predominant. This results in S-shaped form of 0.5 isochlor and resulting large seawater intrusion at the bottom of the aquifer for higher values of d_L.

It is observed from the **fig.-6-c** for Set-C that as the value of d_L varies from 0.1 to 3.0, the shape of 0.5 isochlor remains vertical with more intrusion in all the layers as compared to Set-B. For the d_L equal to 5.0 and higher, intrusion in layer-1 and layer-2 reduces to some extent and increases at the bottom of the layer-3. Thus when the aquifer depths are much larger compared to the longitudinal dispersivity values, 0.5 isochlor remains almost vertical. It is thus inferred that for

the same values of d_L seawater intrusion in single layer coastal aquifer (i.e. for Set-A) and multilayered aquifer for Set-B values is almost same. This study shows that higher values of d_L has significant influence on seawater intrusion for Set-B, whereas it is very meager for Set-C values.

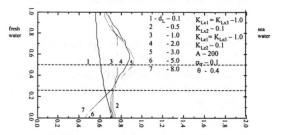

Figure 6-c Effect of parameter d_L on seawater intrusion (0.5 isochlor)

Conclusion

Seawater intrusion problem with velocity dependent dispersion case was formulated and solutions were obtained by Galerkin's finite element method using triangular elements for single and multi-layered confined coastal aquifer. Concentration distribution in the confined coastal aquifer was determined for various conditions of the field parameters within the practical range. The effect of different field parameters on seawater intrusion was investigated.

Following conclusions can be drawn from the present study:

1. Changes in non-dimensional parameter, K_{Lxx} and K_{Lzz} which is the ratio of hydraulic conductivity in horizontal to vertical direction have been found to have negligible influence on the extent of seawater intrusion for single layer confined coastal aquifer. The non-dimensional parameters K_{Lx1}, K_{Lx2}, K_{Lz1} and K_{Lz2} have significant influence on the seawater intrusion for multilayered confined coastal aquifer. Higher is the influence for smaller values of non-dimensional hydraulic conductivity for layer-2. It is observed that for layered aquifer with less hydraulic conductivity for sandwiched layer (Set-C) of the same domain, the

219

seawater intrudes more in all the layers.

2. 'A' is inverse of upstream hydraulic gradient. Seawater intrudes further for higher values of parameter A for Set-B. For Set-C seawater intrusion is more but the influence of 'A' is negligible.

3. The non-dimensional parameter α_T which is the ratio of transverse to longitudinal dispersivity does not have significant influence on the seawater intrusion.

4. For the values of non-dimensional parameter d_L, (which is a ratio of aquifer thickness to longitudinal dispersivity) equal to 1.0 or more, cyclic flow at the aquifer bottom becomes more prominent resulting in increases of seawater intrusion at the bottom and decreases at the top of the aquifer. Whereas, it can be concluded that for Set-B, values of non-dimensional field parameter d_L, upto 5.0 there is a considerable reduction in seawater intrusion in all layers. Whereas, for $d_L = 5.0$ and more, cyclic flow at the aquifer bottom becomes more prominent resulting in the increases in seawater intrusion at the bottom and decreases at the top of the aquifer.

Seawater intrusion for Set-C values is very high as compared to Set-B values, however the effect of d_L on intrusion where it varies from 0.1 to 8.0 is negligible.

Nomenclature

A	-	Inverse of Piezometric gradient
C	-	Concentration of Seawater
d	-	Depth of aquifer
d_L	-	Ratio of aquifer depth to longitudinal dispersivity of the aquifer
D_{xx}, D_{zz}, D_{zx}	-	Components of dispersion tensor
e_j	-	Component of gravitational unit vector
h	-	Equivalent freshwater head
k	-	Coefficient of permeability (Intrinsic permeability)

K	-	Hydraulic conductivity
K_{Lxx}, K_{Lzz}	-	Components of hydraulic conductivity tensor
L	-	Length of aquifer
n_x, n_z	-	Direction cosines
p	-	Hydrostatic pressure
q_x, q_z	-	Darcy velocity in x, z directions respectively
V_x, V_z	-	Seepage velocity in x and z directions
ρ	-	Density of mixed fluid
ρ_f	-	Freshwater density
ρ_s	-	Seawater density
ε	-	Density different ratio
μ	-	Dynamic viscosity of fluid
θ	-	Porosity
α_L	-	Longitudinal dispersivity
α_T	-	Transverse dispersivity
N_i, N_j, N_k	-	Interpolation functions

References

Bear, J., 1979. Hydraulics of ground water, McGraw-Hill, New York, 569 pp

Henry. H.R., 1964. Effect of dispersion on salt encroachment in coastal aquifers, In: Sea Water in coastal Aquifers, U.S. Geol. Surv., Water-Supply Pap. 1613-C, 70-84.

Lee, C.H. and Cheng, R.T.-Sh., 1974. On seawater encroachment in coastal aquifer', Water Resour. Res., Vol. 10, No. 5, 1039-1043.

Pandit A. and Anand S.C., 1984. Ground water flow and mass transport by finite elements - A parametric study: Proc. 5th Int. Conf. on Finite Elements in Water Resources, (J.B. Liable, C.A. Brebbia, W. Gray, G. Pinder eds) Burlington, Vermount, USA, pp 363-381.

Pinder, G.F. and Cooper, H.H., Jr., 1970. A numerical technique for calculating the transient position of the saltwater front, Water Resour. Res., Vol. 6, No. 3, 875-882.

COMMUNITY MANAGEMENT

Intl. Conf. on Sustainable Development and Management of Groundwater Resources
in Semi-Arid Region with Special Reference to Hard Rock, (IGC 2002),
M. Thangarajan, S.N. Rai & V.S. Singh (Eds.)

Ground water protection strategy in Ghaziabad urban area, Uttar Pradesh - a case study

A. Dey and A. N. Bhowmick
Central ground water authority, New Delhi

Abstract

. Ground water quality issues are increasingly emerging as one of the major environmental concern in urban India, consequent upon accelerated growth of anthropogenic activities. The menace of quality deterioration being a serious challenge requires abatement solutions on long term basis through implementation of efficient ground water quality management strategies. To illustrate the technological input in formulation of ground water protection strategies from hydrogeological perspectives, a case study of ground water quality characterization and pollution vulnerability in Ghaziabad urban area has been considered. Central Ground Water Authority has restricted ground water withdrawal by notifying the area.

The deliberations in the present paper includes the assessment of ambient hydrogeological regime of the area, characterization of aquifer vulnerability and suggested protection strategies to mitigate the pollution problems in the area on long term
Basis.

Keywords: Groundwater quality, DRASTIC index and AVI methods

Introduction

Accelerated growth in urbanization, agricultural activities and rapid industrialization in India have created various impacts on environment in terms of ground water quality impairment. Unscientific disposal of Municipal and domestic wastes, industrial effluents and application of excessive fertilizers and pesticides have been predominantly responsible for deterioration in ground water quality. An effective long term management policy planning needs to be evolved to combat and mitigate this menacing environmental degradation. To illustrate the technological input in formulation of ground water protection perspective, a case example of ground water quality characterization and pollution vulnerability in Ghaziabad Urban Area, Uttar Pradesh, India has been considered. Ghaziabad, Uttar Pradesh one of the leading industrial areas in Northern India, had been notified by Central Ground Water Authority in 1999, in view of the depletion of ground water resources and deterioration in ground water quality. The deliberations in the present paper include the assessment of hydrogeological regime, hydrochemical characteristics of ground water, aquifer characteristics and vulnerability in the area. An attempt has been made to evolve and suggest protection strategies to mitigate the ground water problems in the area on long-term basis. The authors believe that the deliberations of this case example will be of immense use for the policy planners in formulation of site specific protection strategies under similar hydrogeological environments.

Study area

The study area comprised of Ghaziabad Urban area (Municipal Corporation area Limit), Ghaziabad district, Uttar Pradesh, India and is located between north latitude $28^0 31' 30''$ and $28^0 42' 30''$ and east longitudes $77^0 19' 00''$ and $77^0 28' 00''$ and encompasses an area of 7000 Ha. The area has overall flat topography with average elevation of 210m above mean sea level, and is drained by south flowing Hindon river. The normal annual rainfall in the area is 527.5 mm. The location map of the study area is presented in fig 1.

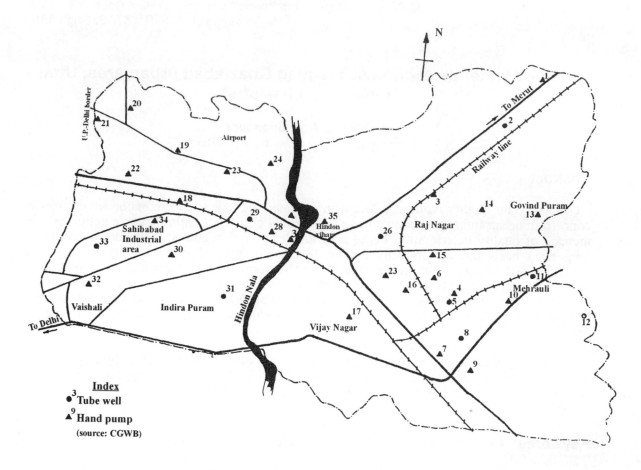

Figure 1: Location map of the study area

Hydrogeological framework

Hydrogeologically, the study area is underlain by moderately thick quaternary unconsolidated alluvial formations (consisting primarily sand and clays) with thickness varying from about 100m to 300m. This unconsolidated formation overlies Precambrian Meta-sediments belonging to Aravalli Super-group. Hydrogeological studies aided with exploratory drilling undertaken by Central Ground Water Board (CGW) in and around the study area reveals the existence of two distinct hydrogeological environments in the study area, on either side of river Hindon which flows along fault line, the eastern (cis Hindon) and western (trans Hindon) parts representing down thrown and up thrown blocks respectively. Hydrogeologically, there exists a single potential aquifer in the trans Hindon area down to depth of 60 to 80 m

bgl and, in Cis- Hindon region, a second group of aquifer exists below a depth of 120m bgl. The quality of ground water in Trans- Hindon area is a serious problem. By and large in the entire Trans-Hindon region ground water below 30 m depth is saline and upconing of saline water interface is caused due to over-development of ground water. The average depth to water level in the area is in the range of 3 to 15 m bgl. The general flow direction of ground water is southerly. The long term water level behaviour as revealed from water level data of National Hydrograph stations, being monitored by Central Ground Water Board indicates decline in pre-monsoon water level in the range of 0.37 to 0.77m on annual basis. The post-monsoon water level behaviour being erratic, no conclusive trend could emerge. The ground water resources of the study area (falling in Loni Block) is estimated to be 9.48 MCM/yr on prorata basis. Taking into account the present

224

withdrawal to the tune of 33.9 MCM/yr of ground water for public and industrial uses in the study area, the stages of development works out to be 358%, which is alarming.

Chemical quality of ground water

A mobile chemical laboratory was deputed by Central Ground Water Authority from 17th –22nd January, 2001 as a follow up action, consequent upon declaration of Ghaziabad Urban area as 'Notified Area', in order to undertake rapid assessment of ambient ground water quality in the area. In all 36 ground water samples, in addition to routine chemical analyses, trace element analyses had also been undertaken to have a comprehensive idea about the hydrogeochemical regime in the study area. The details of analysis are presented in table- 1 and 2.

It is observed from the above tables that concentrations of various parameters in the study area are by and large within the permissible limits of BIS 10500, except in western part of the area where certain quality problems have been observed. From classification point of view (US Soil Salinity Laboratory), it is observed that out of 36 samples, about 53% belong to C3 (brackish) followed by 25% of C2 class (relatively fresh). The saline and very saline water (C4 and C5) are observed in 22 % of samples, representing western part of the study area. The heavy metal concentration was also found to be within permissible limit in the study area. However, in western part, 0.11 percents of samples having concentration of Mn and Pb and 0.06% of samples having Fe beyond permissible limit have been observed.

Assessment of vulnerability

Vulnerability is defined as an intrinsic property of ground water system that depends on the severity of that system to human and /or natural impacts. The assessment of ground water vulnerability to contamination, salinisation and depletion has become decision making tool for assisting planners in minimizing adverse effect on ground water development. There are number of methods available for characterizing aquifer vulnerability based on local hydrogeologic setting, data availability and intended objectives of the projects. Two methods currently in wide use are AVI (Aquifer Vulnerability Index) developed by Prairie Provinces Water Board (Van Stemproort et al, 1992), and DRASTIC, developed by US EPA (Allen et al, 1987) and have been used for the present study.

1. Aquifer Vulnerability Index (AVI)

Aquifer Vulnerability Index (AVI) quantifies vulnerability by the hydraulic resistance (c) to the vertical flow of water through the geologic sediments above the aquifer. Hydraulic resistance is calculated from the thickness (d) of each sedimentary layer and the hydraulic conductivity(K) of each of the layer and is expressed as follow

Hydraulic resistance,

$$c= \Sigma(d_i /K_i) \qquad (1)$$

where, i= 1,2, …,n for n layer

In the present study saturated hydraulic conductivity values (K_{sat}) had been assigned to sedimentary layers as reported in the well records of Exploration undertaken by Central Ground Water Board in and around the area. The hydraulic resistance (c) has the dimension of time (eg in years) and represent the flux time per unit gradient for water flowing down ward through the various sedimentary layers to the aquifer. The lower the value of c, greater is the vulnerability of aquifer. The following classification as per AVI is shown in table- 3. The aquifer vulnerability indices of the surficial aquifer in the study area have been estimated in respect of specific locations and presented in table-4. As revealed from the above table that the AVI values in the study area varies from 0.38 to 6.26 days, with hydraulic conductivity values varying from 16.48 to 66.67 m/day. Which is indicative of extremely high vulnerability of aquifer in the study area.

Table 1: Results of chemical analysis of groundwater samples collected from study area

S.No.	Location	Source	Date	pH	EC	TDS	ALK	Cl	SO4	NO3	F	PO4	Ca	Mg	Na	K	SiO2	TH	RSC	SAR
1.	DUHAI	H.P.	17-01-01	8.20	570	342	190	7	60	25	0.50	Nil	70	9	28	8	36	210	0.4	0.8
2.	MORTA	TW	17-01-01	7.64	760	456	285	14	60	7	1.14	0.30	70	17	50	18	36	254	0.8	1.4
3.	BUDH VIHAR RAJ NAGAR	HP	17-01-01	7.56	1300	780	465	42	80	36	0.78	Nil	60	6	170	24	37	225	5.8	5.6
4.	KAVI NAGAR IND. AREA	HP	17-01-01	7.50	735	441	265	11	70	13	0.48	Nil	62	19	48	20	36	235	0.6	1.4
5.	GD MILL KAVI NAGAR IND AREA	TW	17-01-01	8.03	570	342	190	14	60	13	0.65	Nil	52	21	20	14	33	215	0.5	0.6
6.	BS IND. AREA ROAD	HP	18-01-01	7.63	1720	1032	455	191	110	9	1.05	0.45	46	16	260	80	32	180	5.5	8.4
7.	KL STEEL LTD. LALKAUN	HP	18-01-01	7.60	1900	1140	380	184	280	3	0.60	0.02	210	63	60	20	35	786	-8.1	0.9
8.	S.ALLOYS LTD. BS IND.AREA LALKUAN	T.W.	18.01.01	7.61	750	450	245	22	80	10	0.60	NIL	60	21	50	18	38	235	0.2	1.4
9.	VILL. SHAHPUR BAMHETA	H.P.	18.01.01	7.90	671	403	230	14	70	10	0.48	NIL	80	9	30	18	37	235	-0.1	0.9
10.	SHIV MANDIR MEHRAULI	H.P.	18.01.01	7.93	1780	1068	235	245	140	183	0.72	0.44	122	46	130	76	37	495	-5.2	2.5
11.	UDYOG KUNJ IND. AREA	T.W.	18.01.01	7.86	550	330	160	7	50	6	0.64	NIL	60	7	28	18	38	180	-0.4	0.9
12.	VILL. BAYANA	T.W.	18.01.01	7.94	900	540	285	57	50	30	0.61	NIL	90	30	32	24	40	350	-1.3	0.7
13.	J BLOCK MKT GOVINDPURAM	H.P.	18.01.01	7.80	750	450	210	7	140	4	0.73	NIL	88	15	30	18	36	280	-1.4	0.8
14.	VILL RAIS PUR	H.P.	19.01.01	7.64	195	549	230	57	80	71	1.31	0.21	66	19	80	20	38	245	-0.3	2.2
15.	OPP.D.M. OFICE KAVINAGAR	H.P.	19.01.01	7.61	1250	750	400	64	110	2	0.71	NIL	74	16	150	36	41	250	3.0	4.1
16.	OPP. GANESH HOSPITL N NAGAR	H.P.	19.01.01	7.50	1035	621	315	64	80	23	0.86	Nil	66	19	70	84	40	245	1.4	2.0
17.	DUNDAHERA VIJAYNAGAR	H.P.	19.01.01	7.52	1525	915	475	99	120	23	1.21	Nil	48	19	190	110	37	200	5.5	5.9
18.	RLY STATION SAHIBABAD	H.P.	19.01.01	7.28	1960	1176	190	436	120	43	0.60	Nil	166	44	110	100	32	600	-8.2	2.0
19.	LONI ROAD	H.P.	19.01.01	7.21	3600	2160	740	557	220	37	0.79	Nil	140	67	460	110	46	625	2.3	80

Continue

No.	Location	Type	Date	pH	EC	TDS														
20.	OPP. MATA MANDIR BHOPURA	H.P.	19.01.01	7.09	3810	2286	370	688	240	380	0.65	Nil	280	53	380	112	39	921	-11.0	5.5
21.	NASBANDI COLONY RAJIVE NAGAR	H.P.	19.01.01	7.13	4010	2406	570	858	180	20	0.71	Nil	168	56	540	120	35	650	-1.6	9.2
22.	IQBAL MKT BRINDAVAN GARDEN	H.P.	19.01.01	7.26	1780	1068	235	330	160	23	0.50	Nil	150	35	120	76	41	520	-5.7	2.3
23.	D BLOCK LAJPAT NAGAR	H.P.	19.01.01	7.30	3760	2256	205	801	320	2	0.72	Nil	172	50	480	118	35	635	-8.6	8.3
24.	VILL. KARHERA MOHAN NAGAR	H.P.	19.01.01	8.07	1300	780	210	184	150	9	3.70	Nil	24	13	200	64	25	115	1.9	8.2
25.	GOMTI TOWER KALLUPURA	H.P.	20.01.01	7.80	1210	726	205	71	100	Nil	0.57	Nil	98	23	88	40	42	340	-2.7	2.1
26.	NEAR BRIDGE LOHIA NAGAR	T.W.	20.01.01	7.73	1070	642	305	42	120	34	1.01	Nil	70	25	90	40	39	280	0.5	2.3
27.	SHIV MANDIR NEAR HINDON RIVER	H.P.	20.01.01	7.76	500	300	105	7	100	35	0.40	Nil	48	11	30	12	29	165	-1.2	1.0
28.	BUS STAND ARTHALA	H.P.	20.01.01	7.48	1890	1134	230	401	120	1	0.86	0.33	62	22	260	82	32	245	-0.3	7.2
29.	BANSAL CARBON ANAND IND. ESTATE	T.W.	20.01.01	7.15	4350	2610	695	745	340	61	0.46	Nil	244	131	380	128	50	1151	0.6	6.4
30.	PRAHALAD GARHI	H.P.	20.01.01	7.50	1750	1050	190	401	110	Nil	0.48	Nil	62	36	180	96	31	305	-2.3	4.5
31.	VILL. KANAWANI	T.W.	20.01.01	7.78	545	327	95	14	140	Nil	0.37	Nil	60	6	32	18	28	175	-1.6	1.1
32.	AMBEDKAR PARK VAISHALI	H.P.	20.01.01	7.10	5150	3090	135	1298	540	55	0.23	Nil	178	74	760	120	15	751	-12.3	12.1
33.	PARA PRODUCT SAHIBABAD	T.W.	20.01.01	7.27	3150	1890	285	730	190	70	0.45	Nil	172	73	300	118	33	730	-8.9	4.8
34.	VILL. JHANDAPUR	H.P.	20.01.01	7.05	4220	2532	325	1046	280	18	0.38	Nil	242	98	420	120	39	1011	-13.7	5.8
35.	LK FABRICATOR, 7, SEWA NAGAR	H.P.	20.01.01	7.53	1400	840	400	128	400	132	0.78	Nil	104	29	90	84	38	380	0.4	2.0
36.	OPP. DRAGAH GT ROAD ARTHALA	H.P.	20.01.01	7.25	2000	1200	235	429	235	88	1.08	Nil	140	44	150	96	36	530	-5.9	2.8
	Drinking Water Specifications	Desirable limits		6.50	NG	500	200	250	200	45	1.00	NG	75	30	NG	NG	NG	300		
	BIS:IS:10500,1991	Permissible limit		8.50	NG	2000	600	1000	400	100	1.50	NG	200	100	NG	NG	NG	600		

EC- Electrical Conductance in µS/cm at 25°C, ALK- Alkalinity as CACO₃, TH- Total Hardness as CACO₃, TDS- By Calculation and NG-No Guidelines

227

Table 2: Results of trace elements in groundwater samples collected from study area

S. No.	LOCATION	SOURCE	DATE	Cd	Cr	Cu	Fe	Mn	Nl	Pb	Zn
				<..........................Conc. In mg/l.......................>							
1.	DUHAI	H.P.	17.01.01	N.D	N.D	0.050	0.452	0.160	N.D.	0.030	0.400
2.	MORTA	T.W.	17.01.01	N.D	N.D	0.004	0.043	0.076	N.D	0.020	0.021
3.	BUDH VIHAR RAJ NAGAR	H.P.	17.01.01	N.D	N.D	0.025	0.030	0.011	N.D	0.056	1.290
4.	KAVINAGAR IND.AREA	H.P.	17.01.01	N.D	N.D	0.006	0.035	0.054	0.003	0.020	0.200
5.	GD MILL KAVINAGAR IND.AREA	T.W.	17.01.01	N.D	0.003	0.005	0.004	0.061	0.003	0.016	0.015
6.	BS IND. AREA ROAD	H.P.	18.01.01	N.D	0.050	0.800	0.400	0.200	0.002	0.072	2.740
7.	KL STEEL LTD., LALKUAN	H.P.	18.01.01	N.D	N.D	0.008	0.400	0.280	0.013	0.041	1.250
8.	S.ALLOYS LTD. BS IND.AREA LALKUAN	T.W.	18.01.01	N.D	N.D	0.013	0.100	0.100	0.004	0.021	0.012
9.	VILL. SHAHPUR BAMHETA	H.P.	18.01.01	N.D	N.D	0.008	0.090	0.072	0.003	0.0021	0.450
10.	SHIV MANDIR MEHRAULI	H.P.	18.01.01	N.D	N.D	0.016	0.150	0.220	0.002	0.030	0.130
11.	UDYOG KUNJ IND. AREA	T.W.	18.01.01	N.D	N.D	0.006	0.073	0.050	N.D.	0.030	0.031
12.	VILL. BAYANA	T.W.	18.01.01	N.D	N.D	0.001	0.060	0.120	0.001	0.022	0.025
13.	J BLOCK MKT GOVINDPURAM	H.P.	18.01.01	N.D	N.D	0.015	0.400	0.040	0.003	0.033	3.700
14.	VILL RAIS PUR	H.P.	19.01.01	N.D	N.D	0.020	0.090	0.100	0.100	0.023	0.480
15.	OPP.D.M. OFICE KAVINAGAR	H.P.	19.01.01	N.D	N.D	0.021	0.210	0.031	0.090	0.030	0.060
16.	OPP. GANESH HOSPITL N NAGAR	H.P.	19.01.01	N.D	N.D	0.035	0.200	0.070	0.011	0.025	0.520

17.	DUNDAHE RA VIJAYNAG AR	H.P.	19.01. 01	N.D	N.D	0.024	0.210	0.074	0.100	0.033	1.600
18.	RLY STATION SAHIBABA D	H.P.	19.01. 01	N.D	N.D	0.040	1.020	0.210	0.100	0.034	0.130
19.	LONI ROAD PASONDA	H.P.	19.01. 01	N.D	N.D	0.016	0.300	0.220	0.020	0.035	0.080
20.	OPP.MATA MANDIR BHOPURA	H.P.	19.01. 01	N.D	N.D	0.020	0.500	0.240	0.002	0.026	0.400
21.	NASBANDI COLONY RAJIVE NAGAR	H.P.	19.01. 01	N.D	N.D	0.013	0.400	0.350	0.002	0.030	0.031
22.	IQBAL MKT BRINDAVA N GARDEN	H.P.	19.01. 01	N.D	N.D	0.004	0.130	0.200	0.007	0.026	2.540
23.	D BLOCK LAJPAT NAGAR	H.P.	19.01. 01	N.D	N.D	1.440	0.400	0.040	0.034	0.120	2.210
24.	VILL. KARHERA MOHAN NAGAR	H.P.	19.01. 01	N.D	N.D	0.005	0.011	0.070	0.013	0.002	0.240
25.	GOMTI TOWER KALLUPU RA	H.P.	20.01. 01	N.D	N.D	0.006	0.012	0.200	0.006	0.022	0.200
26.	NEAR BRIDGE LOHIA NAGAR	T.W.	20.01. 01	N.D	N.D	0.004	0.030	0.081	0.004	0.016	0.170
27.	SHIV MANDIR NEAR HINDON RIVER	H.P.	20.01. 01	N.D	N.D	0.011	0.240	0.093	0.005	0.016	1.230
28.	BUS STAND ARTHALA	H.P.	20.01. 01	N.D	N.D	0.014	0.210	0.041	0.001	0.017	0.450

29.	BANSAL CARBON ANAND IND. ESTATE	T.W.	20.01.01	N.D	N.D	0.012	0.065	0.800	0.030	0.040	0.040
30.	PRAHALAD GARHI	H.P.	20.01.01	N.D	N.D	0.020	0.020	0.080	0.010	0.020	0.200
31.	VILL. KANAWAN I	T.W.	20.01.01	N.D	N.D	0.013	0.070	0.280	0.014	0.012	0.020
32.	AMBEDKAR PARK VAISHALI	H.P.	20.01.01	N.D	N.D	0.030	1.740	2.270	0.012	0.050	0.140
33.	PARA PRODUCT SAHIBABAD	T.W.	20.01.01	N.D	N.D	0.020	0.010	0.170	0.010	0.030	0.430
34.	VILL. JHANDAPUR	H.P.	20.01.01	N.D	N.D	0.053	0.500	0.850	0.100	0.043	0.320
35.	LK FABRICATOR, 7, SEWA NAGAR	H.P.	20.01.01	N.D	N.D	0.013	0.010	0.006	0.004	0.020	0.060
36.	OPP. DRAGAH GT ROAD ARTHALA	H.P.	20.01.01	N.D	N.D	0.14	0.210	0.030	0.011	0.023	0.440
Drinking Water Specifications Desirable limits				0.01	0.050	0.050	0.300	0.100	N.G.	0.050	5.000
BIS:IS:10500,1991				N.G.	N.G.	1.500	1.000	0.300	N.G.	N.G.	15.00

Table3: Classification of AVI category based on hydraulic resistance.

Hydraulic resistance (c)	Log c	Vulnerability category
< 10 years	<1	Extremely high vulnerability
10-100 years	1-2	High vulnerability
100-1000 years	2-3	Moderate vulnerability
1000-10000 years	3-4	Low vulnerability
> 10000 years	>4	Extremely low vulnerability

Table-4: Estimation of Aquifer Vulnerability indices of surficial aquifer in the study area

S. No	Location	Thickness Of surficial aquifer(d) in m	Hydraulic Conductivity (Kd) in m/day	Hydraulic Resistance in days	Category of nVulnerability
1.	Sanjay Nagar	61.0	27.87	2.19	Extremely high vulnerability
2.	Shaheed Park	103.15	16.48	6.26	Extremely high vulnerability
3.	Mohan Nagar	41.0	41.46	0.99	Extremely high vulnerability
4.	Tilla Moth	72.0	23.61	3.05	Extremely high vulnerability
5.	Pavi Sadiqpur	25.58	66.67	0.38	Extremely high vulnerability
6.	Ghaziabad	25.25	55.64	0.45	Extremely high vulnerability
7.	Tugalpur Haldena	48.78	36.06	1.35	Extremely high vulnerability
8.	Shahjapur	39.0	64.10	0.61	Extremely high vulnerability

Table 5: Estimatoion of DRASTIC Index in the study area

No	Parameter(1)	Range(2)	Rating(3)	Weightage (4)	Total(3x4)
1.	Depth to water (m)	15-30	7	5	35
2.	Net Recharge (inch)	4-7	6	4	24
3.	Aquifer media	4-9	8	3	24
4.	Soil media	1-10	9	2	18
5.	Topography	2-6	9	1	9
6.	Vadose Zone media	4-8	6	5	30
7.	Hydraulic conductivity (gpd/sqft)	300-700	4	3	12
		TOTAL DRASTIC INDEX			152*

Aquifer with DRASTIC INDEX more than 150 are considered highly vulnerable by USEP

2. DRASTIC index

It is the composite rating of the Depth to Water, net Recharge, Aquifer Media, Soil Media, Topography, Impact of Vadose Zone and the Hydraulic Conductivity of the aquifer and is represented as follows

$$\text{DRASTIC Index} = 5\,D + 4\,R + 3A + 2\,S + I\,T + 5\,I + 3\,C \quad (2)$$

In eq^n .(2) the alphabets represent the corresponding physical parameters and the prefixing numerical represent its respective relative weights. The above parameters have been estimated based on the studies undertaken by Central Ground Water Board. Higher is the DRASTIC index, greater is the vulnerability.

In the present paper the estimation of DRASTIC rating has been computed as per USEPA worksheet and the values were obtained as shown in table-5

231

As discussed earlier, the surficial aquifer in the study area comprised of unconsolidated alluvium consisting of very fine to medium sand with variable clay content. The hydraulic conductivity values are found to be varying from 16.48 to 66.67 m/day. As per ground water regime data, the depth to water table in the area varies. The net recharge based on ad-hoc norms in respect of the study area

Protection strategies

The realization of ground water protection objectives needs to be firmly based on a scientific understanding of ground water behaviour in relation to the threats that the said resource is exposed to. Although the scientific processes of ground water movement and hydrogeochemical interactions in the hydrological cycle are intricately understood, integrating scientific inputs with ground water management options are still to be properly achieved. Ghaziabad urban area (study area) has a semi arid climate with normal annual rainfall of 500mm or so, with significant spatial and temporal variability in precipitation. The scanty rainfall and limited ground water resource availability coupled with indiscriminate withdrawal and vulnerability of aquifers to quality deterioration has resulted in serious concern for the ground water management and protection point of view. With above deliberations under consideration and on establishment of the vulnerability conditions of the aquifers in the study area, the following protection strategies are suggested for the study area:

(1) Inventory, planning and forecasting

It has to be clearly understood that for planning and effecting implementation of urban development activities for a ground water dependent area, the foremost essentiality is the availability of updated information on ground water associated land-use practices and potential pollution causing sources. To achieve this objective, detailed inventories of aquifers should be done including generation of data on their

estimated to be $4.54''$(11.53cm). The aquifer soil and vadose zone media are by and large sandy with admixture of silt and clays. The hydraulic conductivity values as discussed earlier vary from 16.48 to 66.67 m/day. The estimated value of DRASTIC index as per the above rated parameters works out to be 152, which indicates high vulnerability of the shallow aquifer in the study area.

quantitative and qualitative characteristics, their vulnerability to over exploitation and pollution. The evaluation should include data on present situation and future prospects with regard to aquifer use as well as geo-ecological assessment of the impacts of industrial and agricultural activities/facilities on ground water and zoning of ground water protection areas. It has been observed that voluminous data on ground water and related fields have been generated by Central Ground Water Board and State Agencies, which needs to be integrated. Under Hydrology Project generation of micro level database on ground water and related fields are under way, which would definitely serve as potential tool for ground water management in the study area as well. Regarding planning and forecasting, special attention should be accorded to the application of planning tools and forecasting methods for aquifer protection against over exploitation and pollution. Programme for continuous assessment of ground water regime in terms of quality and quantity should be implemented specially for aquifer under risk. In addition to this prospective studies and forecasting of ground water regime, demand and availability of ground water and environmental stress should not only be assessed, be an extrapolation of past trends, but also take into account the anticipated effects of applied foreseen control measures, economic incentives etc for ground water protection. Objective, in particular long term planning, should not only serve the purpose of withdrawal and utilization of ground water but also take care of the protection. Planning should also include the quality forecast of ground water resources for appropriate time horizons, taking into account the ambient pollution and long term pollution on strict implementation of pollution control measures. Suitable ground water models should be developed for multi-variant/multiple forecast of ground water regime particularly for vulnerable aquifers.

(2) Development and implementation of monitoring Programme

Monitoring programmes for ground water protection should be set up and applied. The programmes should include monitoring of ground water regime as well as point and non-point sources of pollution. Ground water monitoring is a critical component of water resource management programme and is defined as an integrated activity for obtaining and evaluating information on the physical, chemical and biological characteristic of ground water in relation to human health and designated ground and surface water users. With accurate information, the current state of ground water resources of Ghaziabad urban area can be better assessed and the water resource protection, conservation and abatement programmes can be run more effectively and the success of management programme can be precisely evaluated. Appropriate ground water modelling techniques with available information can be applied for predicting the future scenario. It has been observed that for the entire Ghaziabad district (2590 sq. km.) monitored through 8 network stations, indicating 324 sq.km coverage per structure and for the study area, it is 70 sq. km. coverage per structure. For a complex hydrogeological scenario, the present density of network stations is inadequate and should be at least 10% of the area i.e. 7 structure for the study area. Periodic monitoring of source of pollution can be taken up in co-ordination with UP State Pollution Control Board and local planning execution agencies.

(3) Implementation of well head protection programmes

Where-ever compatible with national legislation ground water protection zones should be established as a preventive measures for protecting aquifer for present and future abstraction. The purpose of well head protection programmes is to safeguard the public health, safety and welfare of persons in an area by protecting ground water resources from degradation resulting from improper storage, use and disposal of wastes in and around existing and future wellfield and their recharge areas and to promote the economic viability by balancing the protection of ground water with the promotion of economy of the area.

Delineation of well head protection areas (WHPA) is the first step in developing Well-Head Protection Programme. The sub-surface areas which contributes water to a well or well field over a specific time period need to be provided from pollution effect. The delineation of WHPA is based on physical processes governing ground water flow which is also referred as time-of-travel zone (TOT). TOT is the distance traveled by a drop of water through an aquifer to the well or well field in specific period of time. The WHPA size and shape vary in accordance with the hydrogeological features like ground water flow gradient. There are number of methods for delineation purpose like fixed radius, analytical models and numerical flow/ transparent models and hydrogeologic mapping. Modelling studies undertaken by USGS and other agencies have shown 3000 '(1000m) radius circle representing 3 year TOT is optimum for medium to highly transmissive aquifer, which is applicable for the study area. However, this has to be corroborative with hydrogeologic and geomorphic methods, selecting an appropriate WHPA delineation well field requires the consideration of the hydrogeologic setting WHP plans and resources. With an understanding of method eligibility, site specific hydrogeology and management plans, a WHPA delineation well head can be selected that best balances the level of accuracy required with available resources. For fixed radius well head method, it shall depict 1) well head protection area boundary, 2) PWSS pumping well location, 3) Location of significant water withdrawal facilities in the area, 4) Topography of the area, and 5) Well logs of the PWSS pumping wells.

The second step in the Well Head Protection Programme is the inventory of Potential Pollution Sources (PPSI) which includes identification of all current, past and future sources of ground water pollution in the WHPA.

The third step in the Well Head Protection Programme is the source control strategy (SCS) which is developed to minimise the potential for contamination from identified potential pollution sources. In developing a SCS plan, communities may choose contamination of regulatory/non-regulatory management options.

Public Education is the most essential component of successful Protection Programme. The suggested area of activities includes school/public presentation, participation in the local festivals, introduction of ground water curriculum in schools and media coverage.

(4) Setting up of allocation procedure

An appropriate policy compatible with national perspective should be adopted by the local water authorities for preferential allocation of ground water, providing appropriate weight to compatible uses and balancing short term demands with long term objectives in the interest of present and future generations. In allocating ground water, the availability of resources in term of natural replineshable rate and sustainable yield of the basic should be taken into account.

(5) Land use policies and ground water legislation

For framing land use policies, the use of "Hydrological Design Principles" should be adopted which takes into account compatibility of land uses/activities with hydrogeological considerations, location approach etc. The land use policies should take into account the exigency natural ground water replenishable and its protection against pollution and over-exploitation. A co-ordinated approach of ground water management should be adopted wherein the land use policies should be amalgamated/ integrated with other relevant policies of integrated water management. Land use planning should be involved in early stage of development processes. In areas underlain by vulnerable aquifers the ground water protection strategies should carry the decisive weight in land use planning and control. Suitable policies may be evolved and in building bye laws for promoting artificial recharge of aquifer such as roof top rain water harvesting. Site specific legal provisions of ground water management should be formulated and promulgated. Legislation should also contain provisions for its effective implementation involving the mandate and power of relevant authorities. Government rights to control ground water abstraction and area as well as all activities with potential impact on the quantity and quality of ground water resources should be promulgated through legislation. Precise evaluation of ground water characteristics, use and protection should be formulated and integrated into legislation for facilitating implementation of legal provisions for ground water management. It would be necessary to frame up rules covering ground water use rights and for granting permits to uses taking into account the order of priority for allocation of available ground water.

(6) Mass Awareness and Education

Mass awareness and related education programmes are one of the most important proactive approaches considered for efficient implementation of the ground water protection measures. The level of general awareness should be raised in water users through awareness programmes, education, training etc. focussing particularly in respect of availability, possible impacts and vulnerability of ground water resources reservoirs. Active participants of stake-holders in ground water development and management should be involved at all level of planning, decision making and implementation in order to achieve public acceptance of legal and administrative measures which could restrict the freedom of indiscriminate water use.

Conclusion

The impact of accelerated growth in urbanisation and industrialization and manifested declining water table and over exploited conditions and deteriorated water quality are well observed in Ghaziabad Urban Area. The vulnerability assessment of aquifer underlying the study areas reveals alarming vulnerability rating of the aquifers, through DRASTIC & AVI methodologies. To check the twin problem of over-exploitation and vulnerability of aquifers suggested protection strategies are to be adopted. The need for detailed micro-level inventory of aquifers in term of recharge characteristic, withdrawal and vulnerability are to be established. Planning and forecasting of ground water regime, setting up of ground water surveillance and monitoring mechanism are to be adopted. Water allocation procedures, reframing of land use policies compatible with ground water regime, adoption of rain water harvesting in building bye laws and education and involvement of stake holders/users at planning as well as decision making levels are the essential strategies to be observed.

Acknowledgments

The authors are thankful to Dr. D.K. Chadha, Chairman, Central Ground Water Authority for his constant advice and support for the present study. Sincere thanks are due to Sh M.D. Nautiyal, Regional Director, Central Ground Water Board, NWR, Chandigarh for facilitating analysis of heavy metals. Services of Sh. Pankaj Khatri, STA, CGWB and Yashvir Singh, STA, CGWB for undertaking collection and analysis of samples are specially acknowledged. Finally a word of thanks is expressed to Smt. Sonia Kapur, STA, CGWB for rendering assistance to the authors during finalisation of manuscript.

References

Youn Jong Kim, Se-Yeong Hammo (1999). Assessment of the potential for ground water contamination using DRASTIC/EGIS technologies, Cheongju area, South Korea. Hydrogeology Journal 7:227:235.

Allen L, Bennett T, Lehr JH, Petty RJ,and Hackett G,1987, DRASTIC : A Standardised System for evaluating ground water potential using hydrogeologic setting. EPA-600/2-87-035.

Central Ground Water Board,1995, Report on Hydrogeological Framework and Ground Water Resources Potential, Ghaziabad District, UP.

Central Ground Water Board,1997, Report on Status and Problems of drinking water supply in Urban area of Ghaziabad Township ,Field Series Programme- 1996-97.

Central Ground Water Authority,2001, Assessment of Ground Water Quality Concern in India.

United Nations ,1989, Chapter on Ground Water Management PP1-12.

Kaushik, Y.B. and Dey, Arijit ,2001, Rain Water Harvesting- Issues and Options. National Workshop on Sustainable , Chennai.

Dey, Arijit, Chakaraborty, D and Kapur, Sonia ,2000,. Role of Women in Water Management. International Seminar on Women & Water.

Intl. Conf. on Sustainable Development and Management of Groundwater Resources
in Semi-Arid Region with Special Reference to Hard Rock, (IGC 2002),
M. Thangarajan, S.N. Rai & V.S. Singh (Eds.)

Augmentation and management of groundwater resources: participatory watershed management as a policy and strategy

A. K. Sikka

Central Soil , Water Conservation Research and Training Institute, Research Centre
Udhagamandalam - 643 004, Tamil Nadu

Abstract

Over the past three decades, government policies that subsidize credit and rural energy supplies have encouraged rapid development of groundwater resources in India. Problems associated with rapid development are, however, increasing.   Overdraft has become a significant concern in many hard-rock regions. It is causing greater concern due to deficient natural groundwater recharge and this needs to be supplemented by recharge and effective water management.

Harvesting and storage of local runoff and recharging of groundwater aquifers in a framework of integrated land - water development, on a watershed basis with community participation is emerging as a new paradigm due to the recent efforts of both government and non-governmental organizations. An overview of integrated watershed management through participation of local community to promote augmentation and management of groundwater is presented in the paper.

Keywords: Overdraft, hard-rock, groundwater recharge and water management.

Introduction

Over the past three decades competing water demands, increased dependence on groundwater in rainfed/dryland areas and government policies that subsidize credit and rural energy supplies have resulted in a phenomenal growth of groundwater development in India.   This is evident from the fact that groundwater abstraction structures have increased from 4 million in 1951 to nearly 17 million in 1997 (Chadha, 1999). Presently, groundwater contributes about 50 percent of irrigation water, 80 percent of water for domestic use in rural areas and 50 percent of water in urban and industrial areas. In Tamil Nadu, well irrigation accounts for over 60% of the net irrigated area and 11.20 and 14 percent of blocks fall in the category of Dark (draft > 85% of recharge) and overexploited (draft > recharge), respectively.

Rapid pace of groundwater development is resulting in overdraft and water table decline at an alarming rate.   Over exploitation without needed recharge and conservation efforts has resulted in failure of wells, shortage of water supplies, deepening of wells thereby increasing pumping lifts and energy cost and even salinity due to ingression in coastal areas. Overdraft has become a significant concern in many hard rock regions.   In hard-rock areas the specific capacity of wells often declines rapidly after a short period of pumping, indicating limited storage conditions. As a result, depletion often occurs on a seasonal basis even where long-term overdraft problems are not present. It is causing greater concern due to deficient natural groundwater recharge.

The component of rainfall contribution as infiltration to groundwater varies from 3 to 25 percent in different hydrogeological situations (Sinha and Sharma, 1995) and this needs to be supplemented by augmenting recharge and effective water management.   Groundwater resources could be augmented by harnessing non-committed surplus. According to a study conducted by Central Ground Water Board (CGWB), about 214 bcm of surplus monsoon runoff in 20 river basins in the country could be stored as groundwater out of which about 160 bcm is considered to be retrievable (Chadha, 1999).

Sustainability of the resource base is thus critical for meeting an array of basic needs - from health to economic development. Watershed management (WSM) which involves conservation, development and management of water resources at local level or micro scale is gaining due importance and recognition to tackle water resource management including enhancement of groundwater resources.

Participatory watershed management

Watershed management is considered the most appropriate strategy/concept for sustainable development of land and water resources. Participatory approach is more pertinent in the planning and development of watershed management (WSM) programmes, because it is basically the people's programme and the government agency should participate in that as a facilitator. The concept of participatory process is easy to advocate but difficult to practice. This needs active involvement and dedication. Involvement of social organizations, Non-Governmental Organizations (NGOs) and Voluntary Organizations (VOs) is a must to organize and mobilize community support and make this as people's programme and national movement. This whole process may involve following steps.

Regular meetings with the stakeholders in villages to clearly explain the purpose of the programmes, get their feed backs, develop contacts, gather Indigenous Technological Knowledge (ITK), win their confidence and involvement. Appraisal exercises in the village to gather information, diagnose their problems, needs and priorities to arrive at a common outline of watershed development plan.

External implementing agency, governmental or NGO, to provide technical support, act as a facilitator to help them in programme implementation and community organization.

Formation of local People Institution

Institutions for day to day running, management and distribution of benefits and create working capital through revenue generation, people contribution, etc., for repair and maintenance of the works. This will create a self sustaining local institution to take over the activities after withdrawal of the Project.

Watershed management (WSM) involves integrated use and management of rain water, in-situ soil moisture, surface and ground water together with development and management of land, vegetation and crop in a watershed to meet its various needs. Apart from various production and vegetative measures, major components of WSM programmes include a) in-situ soil moisture conservation measures such as bunding, levelling, terracing, vegetative barriers and conservation farming practices; b) surface water development structures such as ponds, tanks, small water harvesting dams, gully control structures, drainage line treatment, diversions, etc.; c) sub-surface/ground water development measures such as percolation ponds/tanks, sub-surface dams/barriers, diaphragm dams, water spreading, sub-surface collection wells, etc.; d) other measures such as roof top collection/rain water cistern, inter-plot water harvesting, etc.; and e) improved water management practices including promotion of micro-irrigation and other water demand management practices.

In general, the above measures of soil-moisture conservation, drainage line treatment and water resource development and management account for about 75 to 80 percent of the total work component in a typical watershed programme. These measures when taken up in a watershed in an integrated manner together with efficient land and production management practices not only augment the water resource but also improve water use efficiency and productivity in the watershed and provide protection against soil erosion and sedimentation of water bodies.

Impact on groundwater

Amongst other benefits, increased groundwater availability as a result of various WSM measures has been reported by different agencies and researchers (Dhruvanarayana et al, 1987; WOTR, Bulletin; Samra, 1997; Gaur, 1998; Sikka et al, 2000). Accurate estimation of groundwater recharge due to WSM measures e.g., soil conservation works, gully plugs, percolation tanks/ponds is hardly available. Generally the groundwater data from a network of observation wells is either not available or lacking in the developmental works. Gore et al (2000) employed water balance model to estimate groundwater recharge and a groundwater flow model in a micro watershed to assess the impact of impounded water in nala bunds on groundwater by simulating water levels with and without water harvesting structures in Wagerwadi watershed in Parbhani district of Maharashtra. The contribution to groundwater recharge was found to vary between 37 to 65% of corresponding surface water impoundment.

In view of the paucity of data, especially the bench mark or base line data in WSM projects, estimation of impact on recharge through simple and surrogate measures has been demonstrated. This involves both, the 'extractive' methods of data collection supported with 'participatory' tools of data/information collection. Impact of WSM on enhancing groundwater availability through indices such as water table rise, enhanced perenniality of wells, recuperation rate, increased irrigated area and crop diversification for different case studies is presented. Largely the Drought Prone Area Programme (DPAP) and Integrated Wasteland Development Programme (IWDP) watersheds of Coimbatore district have been considered in this study.

Water table increase: Rise in water table depth before and after watershed interventions and/or its change over the control well (i.e., outside the influence zone) is generally taken as a direct measure of enhanced groundwater

recharge. Based on the water table data from sample survey of observation wells and piezometers and interview of farmers, it was found that the water level has risen in the range of 0.8 m to 2.0 m in various DPAP watersheds and IWDP Salaiyur watershed in Coimbatore district after WSM interventions. It was also observed that recharge to wells decreased with the distance of wells away from the percolation pond and the influence of these small percolation ponds could be generally observed upto a distance of about 500 m. While in basaltic formations of Central Maharashtra, Rao (1979) found that the area of influence on the average was 1.7 Sq.km with average rise of water level of 2.5 m for percolation tanks (bigger than Coimbatore). Rise in water table as a result of watershed treatment works and percolation pond/tanks at different places

Table1:Groundwater recharge through percolation tanks, water harvesting structures and soil conservation measures

Region	Type of measures	Rise in Water Table (m)
Basaltic formations of Central Maharashtra	Percolation Tank	2.5
Coimbatore district of Tamil Nadu	Percolation Pond	0.8 to 2.0
Chinnatekur, Kurnool, A.P.	Check Dams, bunds and weirs	0.5 to 1.0
Anantapur, A.P.	Water Harvesting Structures	2.0 to 3.0
G.R. Halli, Karnataka	- do -	1.5
Parbhani, M.S.	Nala bund, gully plugs	0.3 to 2.5

in peninsular region as presented in Table-1 does indicate a general rise in water table due to WSM works.

Increased water availability duration in wells

Perenniality of water (i.e., duration of water availability in wells over the year) was taken as a measure to examine as to how the WSM works have helped in increasing perenniality of flow to wells. Data on number of wells having water for different periods in a year (i.e., < 3 months, < 6 months, < 9 months, 12 months) before and after interventions were used in this regard. This fact was generally opined by a number of farmers during the field visit and also ascertained during direct field observations. Figures 1(a) to 1(c) illustrate general increase in perenniality of water in wells for Salaiyur, Arasur and Kattampatty watersheds in Coimbatore district of Tamil Nadu and Mendhwan watershed in Ahemednagar district of Maharashtra. The period of water availability in wells under the influence zone have significantly gone up from 3-6 months before watershed interventions to 9-12 months after watershed interventions.

Increase in recuperation / recharge

Data as regards to duration of pumping hours before well goes dry (or water level depressed to a certain level) and time it takes to recuperate to the same level were collected for sample wells in 13 watersheds of Coimbatore district for before and after execution of WSM works. Recuperation/recharge rate before and after for different watersheds indicated that recharge rate has now increased in the range of 10 to 30%. Before watershed programme, the wells used to go dry after pumping for 0.5-1.5 hours and get recuperated in 38-48 hours. While, after implementation pumping can be done for 1-2.5 hours before well goes dry and it takes 24-36 hours to recuperate. This may be attributed to enhanced groundwater augmentation as a result of WSM.

Increased irrigated area and crop diversification

On the basis of household survey and field observations, it was observed that as a result of enhanced groundwater availability, area under irrigation has increased and farmers have diversified crops also. An increase in area under irrigation by about 7 to 26 percent was observed in different watersheds in Coimbatore.

Table 2: Change in cropping pattern and increased irrigated area of some selected farmers in DPAP watersheds in Coimbatore district

Name of the Farmers	Before		After		Area increase	
	Crops grown	Area in acres	Crops grown	Area in acres	Acres	%
Govindasamy	Flowers & Vegetables	2.0	Banana, Flowers & Vegetables	2.3	0.3	15.0
Palanisamy	Cholam & Vegetables	1.5	Vegetables	1.75	0.25	16.7
Ramasamy	Banana	2.5	Banana & New Coconut (100 Nos)	3.1	0.6	24.0
Ganapathyappan	No crop	Nil	Vegetables & Onion	0.75	0.75	-
Poovatha	Maize	1	Maize & Sugarcane	1.15	0.15	15.0
Rajamani	Onion & Brinjal	1.25	Onion, Banana & Sugarcane	1.40	0.15	12.0
Nellimuthu	Vegetables, Brinjal & Tomato	1.00	Vegetable, Brinjal, Tomato & Banana	1.25	0.25	25.0

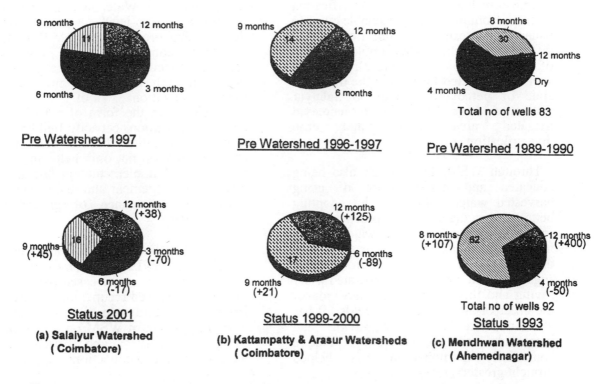

Figure 1: Increased duration of water availability in the wells in selected watersheds after treatment

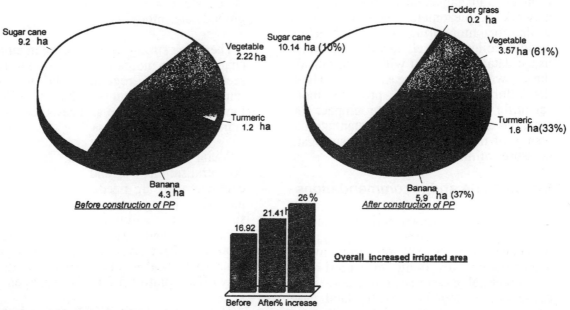

Figure 2: Increased irrigated area and crop diversification in IWDP watershed, Salaiyur, Annur block, Coimbatore district

241

Detailed survey of seven sample farmers in Arasur and Kattampatty watersheds in Coimbatore has shown the diversification of crops from maize and vegetable crops to banana, sugarcane, coconut, flowers and different vegetables after implementation of percolation ponds (Table 2). Sample survey of selected farmers in the influence zone of percolation ponds in Salaiyur watershed have also shown increased irrigated area (26%) and crop diversification (Fig.2).

Through WSM, farmers are also being educated and made aware of using harvested water judiciously by adopting better water management practices e.g., irrigation scheduling and water saving irrigation methods. In Salaiyur watershed, drip irrigation taken up in banana, sugarcane and coconut demonstrated water saving and thereby water conservation of the order of 30% in farmers fields. It has also resulted in labour saving and increased water use efficiency. Active community participation has helped brought greater success.

In and around Salaiyur watershed, Central Soil & Water Conservation Research & Training Institute, Research Centre, Udhagamandalam has set up a network of piezometers and observation wells for monitoring of water table, besides collecting well history and water usage data/information within and outside the watershed areas. These data/information would provide more scientific base to estimate impact on recharge and further help substantiate the results of the above simple and surrogate measures of impact assessment.

Conclusions and recommendations

The Results of various studies pertaining to impact evaluation of integrated watershed management on groundwater augmentation involving structural and non-structural measures of soil and water conservation together with land use management interventions have been quite encouraging. The ridge to valley concept of WSM not only helps to conserve, enhance and manage groundwater but also provides a strategy/concept for integrated planning, development and management of land, water and vegetation for sustainable development. Watershed management involving micro-level water resource development provides multiple benefits to the local community thus minimizing social conflicts. Participatory approach of WSM through community involvement, institutionalization of people's participation in the form of a local level people's institution, capacity building and resource generation through contribution and participation not only helps in proper planning and implementation but also in effective management and maintenance of assets for sustenance of groundwater management.

The results have also demonstrated the use of simple and surrogate measures such as water table rise, increased perenniality of wells, increased recuperation, increased irrigated area and crop diversification for assessing impact of WSM on groundwater using extractive and participatory techniques of data collection:

There is a strong need to have more scientific studies on monitoring and assessment of watershed impact on groundwater including those of isotopes/tracer techniques and groundwater modeling.

There is a strong need to bring in ethical change in the minds of the people to sow seeds of ethics regarding the use and importance of water so that it will no longer be looked upon as a resource which is in plenty and valueless.

Social mobilization, more decentralisation and community empowerment are needed to be explicitly included in government policies for success of community based water harvesting and groundwater recharge projects. Panchayati raj institutions and other such local level people's institutions may offer a platform for this to happen.

The private sector, NGOs and industrial organisations should be maximally integrated in governmental and other institutional initiatives on groundwater augmentation and management.

The encouraging results of watershed management programmes strongly suggest watershed development approach, culminating in total river basin planning, as a policy and strategy in augmentation and management of surface and groundwater with community participation.

Acknowledgement

The author is grateful to Dr. V.N. Sharda, Director, Central Soil & Water Conservation Research & Training Institute, Dehra Dun for his constant encouragement in preparing this paper. The author also acknowledges with thanks the services rendered by Mr. R. Mohanraj, Technical Assistant (T-4) for technical support and Mrs. K. Gnanam, Stenographer for word processing of the manuscript.

References

Chadha, D.K. ,1999, "Groundwater", Chapter 2 of Theme paper on Water : Vision 2050, Indian Water Resources Society, New Delhi, pp. 8-14.

Dhruvanarayana, V.V., Bhardwaj,S.P., Sikka, A.K., Singh, R.P., Sharma, S.N., Vittal, K.P.R.and Das, S.K. 1987, "Watershed Management for Drought Mitigation", Bulletin, ICAR, New Delhi.

Gore, K.P., Pendke, M.S. and Jadhav, S.N. (2000), "Assessment of Ground Water Recharge through Mathematical Modeling in Wagarwadi Watershed in Parbhani District, M.S.", Journal of Soil & Water Conservation, Vol.44, pp.41-47.

Gaur, K.P., Mal, B.C., Pawde, M.N. and Pendke, M.S. ,1998, "Effect of Conservation Measures on Runoff, Soil Loss, Ground Water Recharge and Crop Productivity In Watershed", Journal of Soil and Water Conservation, Vol.42, pp. 68-79.

Rao,, S.S. ,1979, "Effectiveness of percolation tanks as a means of artificial recharge in drought prone areas of Western Maharashtra - A case study", Proc. Int. Symp. Aspects of Droughts, New Delhi, pp. 532-545.

Samra, J.S. ,1997, "Status of Research on Watershed Management" CSWCRTI, Dehra Dun.

Sikka, A.K., Subhash Chand, Madhu, M. and Samra, J.S. ,2000, "Evaluation Study of DPAP Watersheds in Coimbatore District", CSWCRTI, RC, Udhagamandalam.

Sinha, B.P.C., Sharma and Santhosh Kumar Sharma, 1995, "Natural Groundwater Recharge Estimation Methodologies in India", INCOH Secretariat, National Institute of Hydrology, Roorkee.